The
Chromatography
of Hemoglobin

CLINICAL AND BIOCHEMICAL ANALYSIS

A series of monographs and textbooks

EDITOR

Morton K. Schwartz

Chairman, Department of Biochemistry
Memorial Sloan-Kettering Cancer Center
New York, New York

1. Colorimetric and Fluorimetric Analysis of Organic Compounds and Drugs, *M. Pesez and J. Bartos*

2. Normal Values in Clinical Chemistry: Statistical Analysis of Laboratory Data, *Horace F. Martin, Benjamin J. Gudzinowicz, and Herbert Fanger*

3. Continuous Flow Analysis: Theory and Practice, *William B. Furman*

4. Handbook of Enzymatic Methods of Analysis, *George G. Guilbault*

5. Handbook of Radioimmunoassay, *edited by Guy E. Abraham*

6. The Hemoglobinopathies, *Titus H. J. Huisman and J. H. P. Jonxis*

7. Automated Immunoanalysis (in two parts), *edited by Robert F. Ritchie*

8. Computers in the Clinical Laboratory: An Introduction, *E. Clifford Toren, Jr. and Arthur A. Eggert*

9. The Chromatography of Hemoglobin, *Walter A. Schroeder and Titus H. J. Huisman*

ADDITIONAL VOLUMES IN PREPARATION

The Chromatography of Hemoglobin

WALTER A. SCHROEDER
California Institute of Technology
Pasadena, California

TITUS H.J. HUISMAN
Medical College of Georgia
Augusta, Georgia

MARCEL DEKKER, INC.   New York and Basel

Library of Congress Cataloging in Publication Data

Schroeder, Walter Adolph, [date]
   The chromatography of hemoglobin.

   (Clinical and biochemical analysis; v. 9)
   Bibliography: p.
   Includes index.
   1. Hemoglobin--Analysis. 2. Chromatographic
analysis. I. Huisman, Titus Hendrik Jan, joint author.
II. Title. III. Series. [DNLM: 1. Chromatography.
2. Hemoglobins--Analysis. Wl CL654 v. 9 / QY455 S381c]
QP96.5.S34   616.07'561   80-14837
ISBN 0-8247-6941-4

MARCEL DEKKER, INC.
270 Madison Avenue, New York, New York 10016

Current printing (last digit):
10 9 8 7 6 5 4 3 2 1

PRINTED IN THE UNITED STATES OF AMERICA

To Ruth, Glenna, and Rhonda
and Truus, Jeanette, and Door

It has been the wish of the authors of this monograph to compile in one volume a description of the various methods of ion-exchange chromatography that have been developed during the past 20 years for the identification, quantitation, and isolation of normal and abnormal hemoglobins, and to facilitate the choice of a specific type of column chromatography for a specific purpose.

This text contains a detailed description of the art of column chromatography of hemoglobins; it discusses the preparation of the sample, the type of equipment to use, the equilibration of the ion exchanger, the pouring of the column, the application of the sample, gradient systems, the development of chromatogram, and the calculation of the percentages of eluted fractions.

Both macrochromatographic and microchromatographic procedures that make use of different ion-exchange resins and developers are presented. Chromatography on Amberlite IRC-50, CM-Sephadex, CM-cellulose, DEAE-Sephadex, and DEAE-cellulose is discussed with comments about the advantages and disadvantages of each ion exchanger and about the various types of developers for qualitative and quantitative procedures. The descriptions emphasize the applicability of the methods not only to the quantitative determination of major normal and variant hemoglobins but also to that of the minor hemoglobins, such as Hb A_2 and Hb A_{Ic}.

The methods available for the chromatographic isolation and quantitation of fetal hemoglobin are reviewed, and specific methods for the quantitation of Hb Bart's (γ_4) and Hb H (β_4) are described. A section on preparative chromatography for the isolation of larger quantities of selected hemoglobin fractions has a description of methodology with large-sized columns of Amberlite IRC-50, CM-cellulose, CM-Sephadex, DEAE-cellulose, and DEAE-Sephadex that has helpful suggestions for obtaining pure fractions.

This book also contains a chapter about the chromatographic properties of some animal hemoglobins. This section gives examples of the use of different types of ion exchangers and various developers for the separation of hemoglobins of creatures such as the shrimp, rat, goat, cow, deer, and primates.

It has a liberal number of figures which depict chromatographic separations and elution patterns from the application of different ion-exchanger and developer systems to both human and animal hemoglobins. The reader is urged to consult the numerous references when further information on a specific method or gradient system is desired. The Appendix lists many abnormal human hemoglobins that have been quantitated and isolated by one or more chromatographic procedures.

We are grateful to our many friends and co-workers who collaborated at various times in the development of some of the techniques that are described in this monograph. Their names, which may be found in numerous references, will not be listed here. Exceptions, however, are made for Ruth N. Wrightstone, who has our thanks for the preparation of the Appendix and some other sections of the monograph, and for Rhonda De Witt, whose superb typing provided the final camera-ready copy.

Walter A. Schroeder
Titus H. J. Huisman

The Chromatography of Hemoglobin

Hemoglobin fulfills superbly one aspect of the chromatographic (from the Greek χρομα - color and γραφιεν - to write) process: it is highly colored and its progress on the chromatogram (whether satisfactory and satisfying or not) is delightfully apparent. In a sense, it "writes out" its own answer and results by its visibility.

Chromatography has such widespread use today that it is hardly necessary to define it as a dynamic procedure for the separation of substances. As such, it requires that an equilibrium be set up between that portion of substance that is attached to a solid phase and that portion of substance that is dissolved in a flowing fluid phase. Although this goal may be achieved in several configurations, hemoglobin chromatography almost exclusively uses column chromatographic methods in which the solid support is packed in a glass tube through which the aqueous fluid phase is percolated.

Chromatography has been an extremely versatile partner in the study of hemoglobin. Hemoglobin as it comes from the veins of humans or animals is neatly packaged in the red blood cell. Consequently, it is easily separated from exterior plasma proteins. Rarely, however, is the hemoglobin in the red cell a single substance, although in normal and in some abnormal instances a single hemoglobin may account for 90-95% of the mixture. For some experiments, this degree of contamination may not be important; for

others, purification will be vital or, indeed, the contaminant may
be the item of interest.

Chromatography, then, may be applied to hemoglobin in many
ways. It may be used for qualitative identification, for quantitative analysis, or for preparative isolation. It may be used for
the isolation of the hemoglobins in 500 ml blood on a column 10 X
100 cm in dimension (Schroeder and Holmquist, 1966) or for qualitative analysis from 4 µl blood on a column 0.3 X 6 cm (Schroeder
et al., 1975). In column size the factor is about 2 X 10^4 and in
quantity of hemoglobin about 10^5.

Many chromatographic separations of hemoglobins have been described. Some have been designed for separation of simple mixtures
only, whereas others have succeeded in separating very complex mixtures. Some have been used extensively, some only a little. It is
the purpose of this book to supply experimental details for a wide
variety of chromatographic procedures for the hemoglobins, to provide some comments about the application of the methods without
providing an exhaustive survey of the literature, and to point out
some modifications of popular methods. Emphasis will be placed on
human hemoglobins. The viewpoint will be that of the practicing
chromatographer. Simple theoretical considerations will be discussed, but no attempt will be made to provide an insight into
the complex ionic interactions that must go on in a successful
chromatogram.

It is assumed that the reader has more than a superficial
knowledge of hemoglobin. He is presumed to know that

1. Four genetically determined polypeptide chains designated
   α, β, γ, and δ make up the commonly observed normal hemoglobins; ε and ζ chains are present in hemoglobins of the
   early embryonic stage of development.

2. The subunit structural formula $\alpha_2^A \beta_2^A$ represents normal
   adult human hemoglobin A; $\alpha_2^A \gamma_2^F$ represents normal fetal
   human hemoglobin F; and $\alpha_2^A \delta_2^A2$ represents a minor hemo-

globin A_2 (note that all have identical α chains); $\alpha_2\varepsilon_2$, $\zeta_2\varepsilon_2$, and $\zeta_2\gamma_2$ are the subunit formulas for the embryonic hemoglobins that are named hemoglobin Gower-2, hemoglobin Gower-1, and hemoglobin Portland-1, respectively.

3. Point mutations in any one of these chains produce variant hemoglobins such as sickle cell hemoglobin (hemoglobin S or $\alpha_2^A\beta_2^S$) and hemoglobin C ($\alpha_2^A\beta_2^C$).

4. The amino acid sequences of the α, β, γ, and δ chains (but not of the ε and ζ chains) have been determined.*

5. The amino acid substitution in more than 300 variant hemoglobins is known.

6. The β, γ, and δ chains (but not the α and ζ chains) are able to form tetramers such as β_4 (= Hb H), γ_4 (= Hb Bart's), and δ_4.

7. Various blood dyscrasias of known or unknown etiology alter either qualitatively and/or quantitatively the normal and/or variant hemoglobins of a given individual.

More detailed information on any of these and other topics about hemoglobin may be found in many sources, of which the following may be cited: Bunn, Forget, and Ranney (1977); Huisman (1972a); Huisman and Jonxis (1977); Kitchen and Boyer (1974); Lehmann and Huntsman (1974); Ranney (1974); Weatherall (1974, 1976); and Weatherall and Clegg (1972).

Although illustrations of the application of various chromatographic procedures to individual hemoglobins is given throughout the text, the Appendix lists those hemoglobin variants that have been examined chromatographically, provides some data about each one, and cites references where more detailed information is available.

---

*Recently, a partial sequence of the ε chain has been reported (Gale, Clegg, and Huehns, 1979).

Chapter 2
NOMENCLATURE, COMMON FACTORS, AND COMMON PROCEDURES

I. INTRODUCTION

Although the great variation that is possible in the chromatography of hemoglobins is one of the advantages of the methodology, nevertheless there are common aspects to all chromatographic variations. Thus, a sample of blood must be obtained and prepared in some way for chromatography, certain items of equipment are or can be used in all methods, and a qualitative or quantitative assessment must finally be made at the end of the chromatogram. The purpose of this chapter is to describe the nomenclature and abbreviations that will be used, as well as the details of common methodology in order to avoid needless repetition in later sections.

II. NOMENCLATURE

The unabbreviated word "hemoglobin" will be used in a general and all-inclusive sense. Unless otherwise stated, the hemoglobins referred to will be in the form of oxyhemoglobin.

The normal human hemoglobins A, F, and A_2 will be designated as Hb A, Hb F, and Hb A_2 throughout the text. Several subscripts have been used to denote Hb A and Hb F as the main components of adult and fetal blood. Thus, in referring to chromatographic zones, they have been called Hb A_o and Hb A_{II}, and Hb F_o and Hb F_{II}, as well as Hb A_1 in electrophoretic parlance. We will dispense with subscripts for the main components and refer to them as Hb A and Hb F in the text. Some figures, reproduced from other sources, may, however, have the subscripts. For minor components that are posttranslational products of Hb A and Hb F, subscripts will continue to be used.

For the multitudinous variants their given names such as Hb S, Hb D, Hb Bushwick, etc., will denote them.

The distinctions between "microchromatography," "macrochromatography," and "preparative chromatography" depend on column size and are as follows: microchromatography uses columns about 0.5 X 6-20 cm in dimension, macrochromatography about 1 X 20-60 cm, and preparative chromatography much larger, for example, 2 X 25 to 10 X 100 cm.

III.   SAMPLING, ANTICOAGULANTS, AND
       PRESERVATION

The type of anticoagulant, whether heparin, oxalate, EDTA, etc., is
unimportant for samples that have been drawn for hemoglobin chromatography. A 5-10 ml portion of blood is usually adequate for most
analytical studies. However, if microchromatographic methods (Chapter 7) are used, the blood from one or two heparinized microhematocrit tubes from a fingerstick is sufficient. Indeed, for microchromatographic purposes, blood that has been collected on filter paper
is satisfactory.

Unless an unstable hemoglobin is present, blood as such may be
stored for several weeks under refrigeration without undue alteration
of its composition. If a hemoglobin solution has been prepared from
washed cells as described below, it may be stored under refrigeration
for a few days. For storage longer than 1 week, it is advisable to
dialyze against a buffer with cyanide such as the one for any of the
chromatographic procedures to be described in later sections.

The freezing of hemoglobin solutions themselves and storage in
the frozen state may be unsatisfactory because insoluble (presumably
denatured) material may be present after thawing. However, a procedure that VandeBerg and Johnston (1977) devised for preserving
red-cell enzymes also has preserved hemoglobins for more than 6
months in one of the author's laboratory. Thus, 60 g trisodium
citrate dihydrate and 400 ml ethylene glycol is taken to 1 liter
with water; the pH is 8.4. Two volumes of this solution are mixed
with 1 volume of washed, packed cells and stored at -15 to -25°C.
Some samples may not freeze. However, the samples may be thawed
and frozen repeatedly.

Hemoglobin in solution that has been stored in liquid nitrogen
(-70°C) for 2-3 years will retain its chromatographic properties
provided that thawing and refreezing are kept to a minimum (unpublished results). Whether or not freezing and thawing are deleter-
ious may depend upon the type and concentration of ions and the
rapidity of freezing and thawing.

IV. PREPARATION OF HEMOGLOBIN SOLUTIONS

If a hemoglobin solution is required for the chromatographic sample, the following well-known procedure may be used with minor modifications. After centrifuging of the blood and removal of plasma, the cells are washed at least 3 times with several times their volume of isotonic saline (0.85 g dl^{-1} NaCl). Hemolysis is then done by adding water equal to 1-1.5 times the cell volume and organic solvent (preferably carbon tetrachloride) equal to 0.2-0.4 the cell volume. After thorough mixing and intermittent agitation over a period of 10-15 min, the two-phase mixture is centrifuged at 3000 g and at room temperature for 30 min. If a refrigerated centrifuge is available, centrifugation at 4°C is to be preferred. The hemoglobin solution will be above the layer of carbon tetrachloride. Recentrifugation of the hemoglobin solution is often desirable to remove traces of cell debris. If unhemolyzed cells are apparent, a higher ratio of water to cells may be necessary or if the centrifuged solution is not clear, more carbon tetrachloride may be needed. Hemolysates of red cells that contain an unstable hemoglobin are preferably made without the addition of organic solvent; red-cell debris and hemoglobin solution are separated by centrifugation at 30,000 g at 4°C for 60 min.

For many chromatographic separations, the solution so prepared is then dialyzed at 4°C against 100- to 200-fold its volume of an appropriate buffer, which is often the same as the equilibrating solution of the chromatographic column.

Not all chromatographic procedures require such hemoglobin solutions. For microchromatographic methods, it is usually sufficient to prepare the sample from blood to which several volumes of water or a specific solution have been added. On the other hand, for preparative purposes, the use of blood might not be satisfactory because the isolated hemoglobin might be contaminated with plasma proteins.

V. POURING OF COLUMNS

The several ion-exchange media for hemoglobin chromatography (Chapter 4) differ widely in physical properties: whereas the cellulose-based materials and Amberlite IRC-50 form rather dense, white columns, the Sephadex-based materials form translucent columns. Regardless of these differences, columns in general are poured in more or less identical fashion. Thus, the tube in which the column is to be poured is closed at the bottom and equilibrating buffer is layered to a depth of about 5 cm above the support, whether that be a sintered disc, cotton, or glass wool. Then, the tube is filled with thoroughly suspended ion exchanger (usually the slurry has settled ion exchanger and supernatant solution in the ratio of 1:2). As settling occurs and the column reaches a length of about 5 cm, flow is started to speed up the packing. Before all of the first filling has settled, supernatant solution is removed and the tube refilled. This process is repeated until the desired length is attained.

Columns that are poured as described above are, in essence, poured in one section. However, if all the ion exchanger in each filling is allowed to settle completely before more is added, the more finely divided material will be at the top of each section. Both because this more finely divided material often has slightly different properties and because there is an interface between the two layers, a retention of some hemoglobin will sometimes result as a zone passes through the interface, so that striations appear at the interface.

VI. EQUIPMENT

A. Chromatographic Tubes

The most convenient tubes for macrochromatographic columns have a sintered disc to support the column, a ball joint at the bottom so that the flow may be stopped or connection made to the tubing to lead the effluent to some other point, and a socket joint at the top for connection to the supply of developer. It is, of course,

possible to use constricted glass tubing with a support of cotton
or glass wool in the constriction. Such equipment is usually satisfactory for preparative chromatography, but irregularity of flow
through a plug of cotton or glass wool may worsen the separation
of zones on an analytical column.

For microchromatography, columns are packed in Pasteur pipets
or plastic drinking straws (Chapters 7, 8, and 9).

B. Pumps

Ion-exchange columns for hemoglobin chromatography usually have
little back pressure. Consequently, in order to maintain a uniform flow through the column, a peristaltic pump is adequate. In
the authors' experience, Technicon proportioning pumps (Technicon
Instrument Corporation, Tarrytown, N.Y. 10591) have proved most
satisfactory.

Less satisfactory control of flow may be done by adjusting the
hydrostatic head on the column by raising or lowering the vessel
that contains developer. Usually there is a decrease in flow rate
under these conditions as the chromatogram progresses. Control of
flow by hydrostatic head may be awkward when gradients are used.

C. Gradient Vessels

Gradient vessels commonly are constant-volume-mixer type or multivessel type. The constant-volume-mixer type may easily be made
from an Erlenmeyer flask of appropriate size, a separatory funnel,
rubber stopper, and glass tubing. A two-vessel example of the
multivessel type requires glass cylinders with connections and a
stopcock. The apparatus and the type of gradient that is produced
by these systems are discussed in Chapter 4. Numerous gradient
makers are available commercially.

D. Fraction Collectors

Fractions normally need to be collected only from analytical columns on a macrochromatographic scale. Because fraction size
normally is 3-6 ml for columns of this size, most of the many

small fraction collectors now on the market are satisfactory. If
many chromatograms are being run simultaneously, the old-style
fraction collectors with 50 or more tubes per two to four rows may
be useful.

E. Spectrophotometry

Because percentages of the various hemoglobin components are calculated from analytical columns, the exact quantity that is used
is unimportant and, in fact, may vary over an appreciable range
without untoward influence on the separation and the analytical
data. However, a good estimate of the concentration of a hemoglobin solution in oxy- or carbonmonoxy-form may be obtained by this
formula:

$$C = 1.14 \times A_{540} \times f$$

where C is the concentration in milligrams per milliliter, 1.14 is
a constant derived from the extinction coefficient, A_{540} is the
absorbance at 540 nm of a 1-cm length of solution, and f is the
dilution factor between the original solution and a portion that
has been diluted to provide a solution with readable absorbance.

Because the extinction coefficient of the Soret band at 415 nm
for oxy- or carbonmonoxyhemoglobin is approximately 10-fold that at
540 nm, absorbance readings from the effluent of a chromatogram are
usually made at 415 nm for greater accuracy in the quantitation of
minor components. Such readings may be made on the individual
fractions with a large number of commercially available spectro-
photometers some of which have accessories to provide automatic
printout of absorbances. Because 200 or more fractions may well
be collected in the course of an analytical chromatogram, it is
convenient to use a commercially available apparatus in which the
spectrophotometer cell is automatically filled and emptied with
consecutive fractions.

The course of the chromatogram may also be monitored by passing the effluent through an absorbance monitor. Whether such an

instrument is useful will depend on its ability to cope with extremes of absorbance. Because the difference in content of major and minor components in chromatograms may range over a factor of 25-75, such a monitor may only be qualitatively useful unless the instrument automatically adjusts its absorbance range.

VII. CALCULATIONS

For accurate quantitative analyses, the volumes of the fractions need to be uniform (± 2%) (and this usually is true if a pump is used) or they must be diluted to equal volumes. If nonuniform fractions are likely to result from the experimental arrangement, it is then convenient to use calibrated test tubes for collection, to collect fractions slightly smaller than the calibrated volume, and then to dilute to volume. Some fractions will usually be so concentrated that further dilution will be necessary to come within the absorbance range of the spectrophotometer. After absorbance readings have been made on all fractions from a chromatogram, the absorbances of fractions for all peaks are totaled individually. If fractions have been diluted, the effect of such dilutions must be accounted for by multiplying the absorbance of each diluted fraction by a factor which is the ratio of the volume of the diluted fraction to the volume of the undiluted fractions. When the absorbances have been summed, the percentage of component X of N components is

$$\% \ X = \frac{\Sigma A_x \times 100}{\Sigma A_1 + \Sigma A_2 + \cdots + \Sigma A_n}$$

where ΣA_1, etc., is the total absorbance under any peak.

In microchromatographic procedures (Chapter 7, 8, and 9), conditions are so arranged that the separated zones are each in a single fraction, although the volumes of these fractions may be different. In calculating percentages, therefore, the absorbance of component X must be multiplied by the ratio V_x/V_s where V_x is the volume in which X is dissolved and V_s is the smallest volume in which any of the components is dissolved. The general formula, thus, is:

$$\% \ X = \frac{(V_x/V_s)A_x \ X \ 100}{(V_1/V_s)A_1 + (V_2/V_s)A_2 + \cdots + (V_n/V_s)A_n}$$

where the A's are the absorbances of the various components.

VIII.   pH AND CONDUCTANCE MEASUREMENTS

As will be discussed in Chapter 4, Section IV, gradients in pH or ion concentration are often used as developers in hemoglobin chromatography. If that gradient is determined in the effluent fractions, not only will any aberration in the desired gradient be detected, but the pH or ion concentration at the point of emergence of any component will be known. The latter information is of value in attempting to improve separations, in defining conditions for preparative chromatography, etc.

A gradient in pH is easily defined if the pH of every tenth fraction from a chromatogram is determined.

The concentration of NaCl in Tris or bis-Tris is a linear function of conductance in mhos (that is, the reciprocal of resistance in ohms). Consequently, if the conductance of the limiting developers of a gradient is plotted against NaCl concentration, the NaCl gradient is easily defined by determining the conductance of every tenth fraction and interpolating the NaCl concentration from the graph. Whether such a linear relationship of concentration to conductance exists in phosphate solutions or NaCl in phosphate has not been studied.

The osmolality of a solution which is measured by freezing-point depression may also be used to determine the NaCl concentration in the effluent. Thus, after the osmometer has been calibrated with solutions of known osmolality, the NaCl concentration may be calculated from the observed depression of the freezing point in various effluent fractions.

IX. CONCENTRATING HEMOGLOBIN SOLUTIONS

After chromatographic separation, an individual hemoglobin may be present at high dilution in a relatively large volume of solution (for example, 1-2 mg Hb A_2 in 50 ml). Even at perhaps 10-fold that concentration for a main component, the solution needs to be concentrated before other studies such as electrophoresis, oxygen dissociation, etc., can be done. The applicability of various methods of concentrating hemoglobin solutions will depend upon the degree to which unaltered functional capability must be maintained.

A. Evaporation

If functional capability is unimportant (for example, if an amino acid analysis is to be made), it is adequate to dialyze the hemo- globin solution thoroughly to remove buffer ingredients and then to evaporate the water either by lyophilization or by a stream of air.

B. Centrifugation

A mild but burdensome (especially if the volume is large) and some- what expensive method of concentration is high-speed centrifugation (Vinograd and Hutchinson, 1960). Thus, if the dilute solution is centrifuged at 125,000 g at 0-5°C for 16-24 hr in a preparative centrifuge, the hemoglobin will concentrate in a few tenths of a milliliter at the bottom of each centrifuge tube and can readily be removed with a long needle on a syringe.

C. Ion Exchange

Because the degree of fixation of hemoglobin to an ion exchanger is so dependent upon ion concentration and pH (Chapter 3), these conditions in any hemoglobin solution can be altered so that the hemoglobin becomes very strongly fixed upon passage through a col- umn of appropriate ion exchanger. A procedure of this type is described by Schroeder *et al.* (1970).

CM-Sephadex is equilibrated with 0.05 M Tris-maleic acid buffer at pH 6.5 as described in Chapter 5, Section III, and a column 1-3 cm in length is poured. The diameter of the column will depend upon the amount of hemoglobin to be concentrated rather than upon the volume of solution. Thus, a 0.5 X 1-3 cm column will concentrate at least 10 mg hemoglobin from several hundred milliliters of solution. The hemoglobin solution to be concentrated is then diluted with water to reduce the ion concentration to about 0.025 M or, in the case of 0.2 M glycine developer, to about 0.1 M. After addition of one or two drops of 2% KCN solution, the pH is adjusted to 6.5-6.7 and the solution is passed rapidly through the short CM-Sephadex column. When the sample has entered the column, it is washed with a few milliliters of the 0.05 M Tris-maleic acid buffer at pH 6.5. Finally, the fixed hemoglobin is eluted from the column with a minimum volume of 2% KCN. This solution may then be dialyzed against water or a buffer appropriate for the next experiment.

D. Ultrafiltration

Several variations of ultrafiltration may be used. A simple procedure uses dialysis tubing, which is suspended in a filter flask and tightly fixed to a glass tubing that passes through a stopper. The dialysis tubing, glass tubing, and an attached reservoir are then filled with hemoglobin solution. When vacuum is applied to the filter flask, water and small molecules such as buffer ingredients pass through the dialysis membrane, but the hemoglobin is retained and concentrated.

A more elaborate means of ultrafiltration uses commercially available equipment in which a membrane which may have various molecular weight cutoff points is placed in the bottom of a cell. The solution to be concentrated is then placed on the membrane and may be stirred gently while gas pressure (usually N_2) is applied to force water and small molecules but not, for example, hemoglobin, through the membrane. Concentration in this way can be very rapid

because a wide choice of membranes is available. If only a small
volume is to be concentrated, such membranes may be had in the form
of cones and centrifuged to force water and small molecules through
the membrane. Suitable equipment can be obtained from Schleicher &
Schuell, Inc. (Keene, N.H. 03431) and Amicon Corp. (Lexington, Mass.
02173).

Another type of ultrafiltering unit is an immersible filter.
The Millipore Corp. (Bedford, Mass. 05130) supplies Immersible CX
Ultrafiltration Units which are immersed in the solution. When
vacuum is applied, water and small molecules are removed.

Chapter 3

PRINCIPLES OF THE CHROMATOGRAPHY OF HEMOGLOBINS

I. THE TITRATION CURVE

Like all proteins hemoglobin is amphoteric: it may behave either
as an anion or as a cation.   The titration curve of human oxyhemoglobin as depicted in Figure 3.1 shows an isoelectric point at pH
6.75 (Antonini *et al.*, 1965).   As a consequence, hemoglobin will
be an anion at a pH slightly below and above neutrality or a cation
at a pH not far below.   Clearly, as the pH is moved farther and
farther from the isoelectric point in either direction, the positive or negative charge on the molecule will increase and it will
be more cationic or anionic.

II. THE ION-EXCHANGE MATERIAL AND
IONIC CHARACTER

For successful chromatographic separation of hemoglobins, ion-
exchange materials have been used as the solid support.   Because
the chemical nature of these materials will be discussed in Chapter 4, it is enough to note here that ion-exchange materials are
classified as strong and weak cation and anion exchangers.   It is

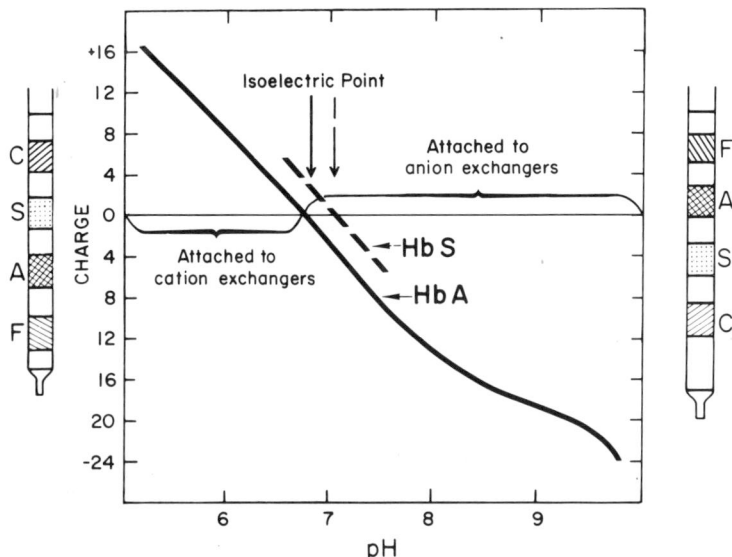

Figure 3.1. Titration curve of human oxyhemoglobin and a correlation with chromatographic behavior. Dashed line represents difference between isoelectric points of Hb A and Hb S. (Adapted from Antonini *et al.*, 1965.)

the weak cation and anion exchangers that are useful in hemoglobin chromatography. The fact that hemoglobin can exist both as an anion or as a cation and that both anion and cation exchangers are available is a great advantage in hemoglobin chromatography. Thus, at a pH below the isoelectric point, the hemoglobin will behave as a cation and, as noted in Figure 3.1, will be attached to cation exchangers. Conversely, the opposite will be the case at a pH above the isoelectric point.

The relevance of this information to the separation of hemoglobin may be seen from this consideration. Pauling *et al.* (1949) in their famous paper on sickle cell anemia noted a difference of 0.22 pH units in the isoelectric points of Hb A and Hb S. This difference is represented by the dashed line of Figure 3.1. Thus, at pHs below the isoelectric point, Hb S will be more cationic than Hb A. Similarly, above the isoelectric point, Hb A will be more

anionic than Hb S. Consequently, we would expect Hb S to move
more slowly than Hb A on a cation-exchange column and more rapidly
on an anion-exchange column. This is depicted in the representations of chromatograms in Figure 3.1. These representations show
relative movement only and are not drawings of actual chromatograms.
The ability to use both cation and anion exchangers for the chromatography of hemoglobins permits one to choose the order in which
the hemoglobins of a given mixture emerge from a chromatographic
column. There is often some advantage in this ability. For example, if the electrophoretically fast-moving Hb N-Baltimore $(\alpha_2\beta_2$
95 Lys → Glu) is to be isolated, it may be useful to choose a cation exchanger in which the Hb N moves first through the column
rather than an anion exchanger on which it is strongly fixed. On
the other hand, if other components can easily be separated from
this Hb N on an anion exchanger, it may be advantageous to carry
out the separation, to remove the top section with the Hb N to another chromatographic tube, and with a strong developer to elute
the Hb N as a concentrated solution.

Pauling *et al.* note that the titration curve is nearly linear
near the isoelectric point and that a change of 1 pH unit alters
the net charge by about 13 charges per molecule. Although charge
is a vital factor in the chromatographic separation of hemoglobins
in a specific set of conditions, it is not the only factor. The
electrophoretic behavior of Hb F in alkaline medium places it between Hb A and Hb S, but chromatographically it does not take this
position (Figure 3.1). The α chains of Hb A, Hb S, and Hb F are
identical; Hb S and Hb A differ in a single amino acid residue at
position 6 of the β chains, where Hb S has valine and Hb A has glu-
tamic acid, but the γ chains of Hb F differ in 39 positions from
the β^A and β^S chains. Thus, other forces that are associated with
the amino acid sequence and the arrangement of side chains of the
protein apparently also influence the chromatographic behavior.
Indeed, variant hemoglobins that electrophoretically behave like
Hb A or Hb F may be chromatographically separable; for example, Hb

Leslie ($\alpha_2\beta_2$ 131 Gln → delete) has the electrophoretic behavior of
Hb F on starch gel at alkaline pH, but readily separates from Hb F
chromatographically (Lutcher *et al.*, 1976).

III. FACTORS THAT INFLUENCE
      CHROMATOGRAPHIC BEHAVIOR

For any substance under a given set of conditions, the chromatographic behavior will be governed by the equilibrium between the
quantity of substance that is attached to the solid phase and that
which is present in the fluid phase. Thus, the equation $K = Q_S/Q_F$
applies where K is a constant, Q_S is the quantity of substance on
a unit weight of solid phase and Q_F is the quantity of substance in
a unit volume of fluid phase. For a substance whose K approaches
infinity (although, in fact, K need not be very large and still
mimic infinity), all or almost all of the material will be attached
to the solid phase. Despite the dynamic aspect of chromatography,
the substance will move very slowly or virtually not at all down a
chromatographic column because so few molecules are present to be
swept away by the fluid phase to set up a new equilibrium with a
new section of solid phase. Conversely, as K approaches zero, the
substance will be swept down the column with little or no attachment to the solid phase. Obviously, an intermediate value of K
rather than either of these extremes is desirable. Equally obvious,
chromatographic separation of substances in a mixture will occur
only if the K's differ. Although sophisticated mathematical exami-
nations of chromatographic behavior have been published, chromatography is still in large part an art. The successful chromatographer
understands how to manipulate the K's in a particular system.

     Of the three factors in any chromatographic separation -- the
substances to be separated, the solid phase, and the fluid phase --
the substances themselves are subject to least manipulation, except
perhaps by derivatization, which may lead to a very different chromatographic system. In hemoglobin chromatography, derivatization
is usually limited to modification of the ligand to the iron of

the heme. Chromatography of deoxyhemoglobin has rarely been attempted, and exchange of carbon monoxide for oxygen leaves the
chromatographic properties unaltered. Oxidation of the iron from
ferrous to ferric state changes chromatographic properties, probably mainly because of the charge difference, but this difference
can for practical purposes be eliminated by adding cyanide to the
medium to form ferrihemoglobin (methemoglobin) cyanide which has
the same charge as oxy- or carbonmonoxyhemoglobin. Thus, derivatization is usually unimportant in the chromatography of hemoglobin,
although, in rare instances, differential derivatization has provided a means for altering the electrophoretic and chromatographic
properties of hemoglobins in a mixture, and thus permitting separation (Brennan, Winterbourn, and Carrell, 1977; Garel *et al.*, 1974).
In vitro incubation of hemoglobin with either glucose or glucose-
6-phosphate modifies its electrophoretic and chromatographic properties (see Chapter 8), whereas treatment of hemoglobin with p-
chloromercuribenzoate (Bucci and Fronticelli, 1965) results in the
attachment of this agent to sulfhydryl (-SH) groups of both α and
β chains and in dissociation of the protein into single α-PMB and
β-PMB polypeptide chains which can readily be separated by chromatography (Abraham and Huisman, 1977).

As already noted, a choice may be made between a cation and
an anion exchanger as the solid phase for hemoglobin chromatography.
The choice often is immaterial or may be dictated by the particular
objectives of the experiment.

Two variations of the fluid phase are important factors in the
chromatography of hemoglobin. These are the pH and ionic concentration. As the pH is decreased, the cationic character of hemoglobin increases and hemoglobin will be most strongly fixed to a
cation exchanger: its K increases. Conversely, on an anion exchanger, stronger fixation occurs as the pH is increased. Accord-
ingly, it is common to place the sample onto a cation exchanger at
a relatively low pH and onto an anion exchanger at a relatively
high pH in order to ensure that the components will become fixed.

In order to change the K to a value so that the hemoglobin(s) begin to move down the column, the pH may be raised in a cation-exchange system and lowered in an anion-exchange system. The manner in which this is most effectively done will be discussed in more detail in Chapter 4. On the other hand, at a constant pH, a decrease in K may be effected by increase in ionic concentration, whether this be by increasing the concentration of the buffer that is controlling the pH or simply by the addition of sodium chloride. Both pH and ionic concentration may, of course, be varied simultaneously.

Temperature will also influence chromatographic behavior. Usually an increase in temperature will decrease affinity for the solid phase. Much of the effect of temperature may be due to pH change, especially in buffer systems such as Tris (tris-(hydroxy-methyl)-amino-methane), which have a large temperature coefficient of pH.

Chapter 4

CHARACTERISTICS OF ION EXCHANGERS AND DEVELOPERS

I. CATION EXCHANGERS

Strong ion exchangers, such as those with sulfonic acid groups on
a polystyrene backbone, have found no use in hemoglobin chromatography. Rather, three weak ion exchangers, all of which have the
carboxyl group as the active group, have been most effective and
most frequently used. Amberlite IRC-50, a synthetic material, is
a copolymer of 95% methacrylic acid and 5% divinylbenzene. Car-
boxymethyl-cellulose and carboxymethyl-Sephadex are prepared by
the action of a haloacetic acid on cellulose or dextran, respectively. Physically, Amberlite IRC-50 and CM-cellulose are somewhat
similar and form white, rather dense, solid chromatographic columns.
The materials tend not to swell or shrink greatly in fluids of different ionic concentration. Quite the contrary, 1 g dry CM-Sephadex

will imbibe 25-30 times its weight of water and form a translucent, soft column whose surface is easily disturbed.

These ion exchangers differ in the number of titratable groups per unit weight; Amberlite IRC-50, CM-cellulose, and CM-Sephadex have about 10, 1.0, and 4.5 meq per dry gram, respectively. Such a measure is not easily correlated with the ability of the material to achieve a particular separation. Presumably, a more meaningful number would be in terms of milliequivalents per unit volume of packed column. Such information would be difficult to determine accurately because many factors such as packing density, particle size, accessibility of macromolecules instead of small ions to the active site, etc. would influence the number. As an example, one may note that 1 g dry CM-Sephadex becomes 25-30 ml swollen material; consequently, its effective capacity per milliliter will be reduced to about 0.2 meq, whereas in contrast, Amberlite IRC-50 will have about 3.3 meq ml^{-1} in a packed column.

Amberlite IRC-50 is a product of Rohm and Haas (Philadelphia, PA) and is also sold by the Bio-Rad Laboratories (Richmond, CA) as Bio-Rex 70 in various mesh sizes.

CM-Sephadex is produced by Pharmacia (Uppsala, Sweden, and Piscataway, N.J.). As "C-50," it is spheres with a range of diameters from 40 to 120 μ.

Before CM-cellulose became commercially available, experimenters with hemoglobin chromatography used material that they had prepared by the method of Peterson and Sober (1956) in their original description of ion-exchange celluloses. Because of less exact control of procedure than in modern commercial practice, these preparations were more variable in properties, and exact duplication of their results was not always possible. At the present time, CM-cellulose is available from many commercial sources and there is generally good reproducibility from lot to lot. Although expensive, one of the finest commercial preparations is a microgranular, pre-swollen form designated "CM-52" from Whatman, Inc. (Clifton, N.J.).

Because of its preswollen state, equilibration requires a minimum
of effort.

II. ANION EXCHANGERS

The most effective anion exchangers for hemoglobin chromatography
are DEAE-cellulose and DEAE-Sephadex. The DEAE (diethylaminoethyl)
group is combined through the hydroxyl groups of dextran or cellulose to give

$$(C_2H_5)_2\text{-}\overset{\overset{\displaystyle H^+X^-}{|}}{N}\text{-}C_2H_4O\text{-}R$$

where R is the dextran or cellulose moiety.

Most of what has been said in comparing Sephadex- and cellulose-
cation exchangers is equally applicable to these anion exchangers.
DEAE-Sephadex from Pharmacia is designated "A-50" and the microgranular, preswollen DEAE-cellulose from Whatman is called "DE-52."

III. DEVELOPERS

Only a limited number of substances have been used as buffers in
developing solvents for hemoglobin chromatography. Phosphate, because of the fortunate position of its pK values, covers a wide
range of pH and has been used both with cation and anion exchangers.
Tris-hydroxymethylaminomethane (Tris) with pK of 8.3 is suitable in
the basic range with anion exchangers. N,N-Bis-(2-hydroxymethyl)-
imino-tris-(hydroxymethyl)-methane, or bis-Tris, a compound similar
to Tris but with a pK of 6.5, has recently been used in the acid
range with CM-cellulose (Schroeder *et al.*, 1976). The type of ion
in the buffer is not without its effect. Thus, greater concentrations of Tris or bis-Tris are required when they, rather than phosphate, are the buffering ion. Compared to phosphate, Tris or bis-
Tris influences the chromatographic separations -- sometimes for
better and sometimes for worse.

Sodium chloride has been used as a component of developers
for hemoglobin chromatography. It may be used with any of the

above buffers and with both cation and anion exchangers. Most commonly, a gradient of NaCl is used in a buffer at constant pH and buffer concentration. Sodium chloride is effective in promoting the movement and separation of hemoglobins, presumably by providing ions to exchange. However, this is probably too simple a view of the influence of NaCl. Thus, the resulting chromatograms may not be and usually are not identical when a gradient of sodium chloride at a given pH and buffer concentration is compared with a gradient of buffer at that same pH.

Glycine has been a recent addition to components of developers for hemoglobin chromatography (Abraham *et al.*, 1976-1977; Huisman *et al.*, 1975). Under the specific conditions of use, glycine may not be acting as a buffer in this system, because the pH is in the isoelectric range of glycine. The concentration of glycine, which is 0.2 M, is higher than that of other components, and its action may lie in some effect of the dipolar ion. Whatever may be the basis for its action, the practical result is compact zones on a DEAE-cellulose chromatogram that are in great contrast to the diffuse zones on DEAE-Sephadex chromatograms with Tris-HCl gradients.

Allen, Schroeder, and Balog (1958) introduced the use of KCN in developers for hemoglobin chromatograms. Solutions of oxyhemoglobin inevitably contain some ferrihemoglobin which will produce slowly moving extraneous zones on a cation-exchange chromatogram. The addition of KCN converts the ferrihemoglobin to ferrihemoglobin cyanide, which has the same chromatographic behavior as oxyhemoglobin, and thus the extraneous zones are eliminated. Allen, Schroeder, and Balog used 0.01 M KCN (0.65 g liter^{-1}) in their buffers whereas Huisman and Meyering (1960) added only 0.1 g liter^{-1} (0.015 M) in their buffers. The lesser quantity is unquestionably adequate. However, one cannot indiscriminately reduce the amount of KCN in the developers of Allen, Schroeder, and Balog without influencing chromatographic behavior; the quantity of KCN influences both ionic concentration and pH of a given solution.

IV.   DEVELOPER SEQUENCES

A.   Types of Sequence

In general, there are three ways in which chromatographic development may be done: (1) single solution; (2) stepwise change to
solutions of different composition; and (3) gradual, continuous
change in composition of solution.

     When a single solution is used, the column usually is equili-
brated with that solution, the sample is dissolved in that solution,
and that solution alone produces the movement and separation of the
substances. This type of development is most useful if the mixture
is relatively simple and the rates of movement of the zones are not
too disparate. If the mixture is more complex and if the rates of
movement differ widely, it may be difficult to devise conditions
for a successful chromatogram; strongly fixed substances form diffuse zones as they emerge from the column and are contained in
large volumes of effluent. On the other hand, a single developer
can be very useful, if less strongly fixed components emerge fairly
rapidly in the filtrate and separate fractions can be collected,
while the more strongly fixed materials separate but remain on the
column. Subsequent procedure would be to remove those sections of
the column individually, pour each into a separate column, and elute
the substance with another solution. Such a procedure is described
in Chapter 10 and is excellent for preparative experiments.

     Stepwise development is a succession of different solutions
that differ in pH, concentration, composition, or any combination
of these three variables. This type of development is common in
amino acid analysis where a change from buffer at pH 3.25 to pH
4.25 quickly and with good separation brings out amino acids that
would require a large volume of pH 3.25 buffer and result in diffuse zones. If proper conditions can be devised, stepwise development is an excellent means of producing a successful chromatogram.

     Gradient development is really stepwise development in infi-
nitely small steps and relies on a mechanical device to mix two
solutions in such a way that there results a continuous change in

pH, concentration, composition, or any combination of these variables. It is one of the more popular means of development in hemoglobin chromatography and will be discussed in some detail in the following section.

B. Gradient Systems

Gradients of almost any complexity may be designed by use of the proper apparatus. In fact, commercial devices permit a gradient to be designed simply by drawing a profile (LKB Ultragrad). However, such elaborate equipment is not necessary, and hemoglobin chromatography is adequately performed by use of simple constant-volume or multivessel systems.

1. *Constant-Volume Systems*

A constant-volume system uses a *closed* vessel (constant-volume mixer) such as an Erlenmeyer flask with a stopper, which has an inlet just below the stopper and an outlet near the bottom of the flask (Figure 4.1). The outlet goes to the chromatographic column and the inlet is attached to a reservoir. If the mixer is filled with solution 1 and the reservoir with solution 2, withdrawal through the outlet will bring in an equal volume from the reservoir to maintain a constant volume in the mixer. By thorough mixing in the mixer, the effluent from the mixer will have a changing composition that initially is that in the mixer and eventually approaches that in the reservoir. An equation for calculating the nature of the gradient is given in one of the first treatments of the topic of gradient development by Alm *et al.* (1952). This equation is

$$2.303 \; \log_{10} \frac{C_2 - C_1}{C_2 - C} = \frac{v}{V}$$

where C_1 is the concentration of some substance in solution 1 and C_2 its concentration in solution 2. When a volume v has flowed through a mixer of volume V, the effluent concentration is C. The nature of the gradient is depicted in Figure 4.1.

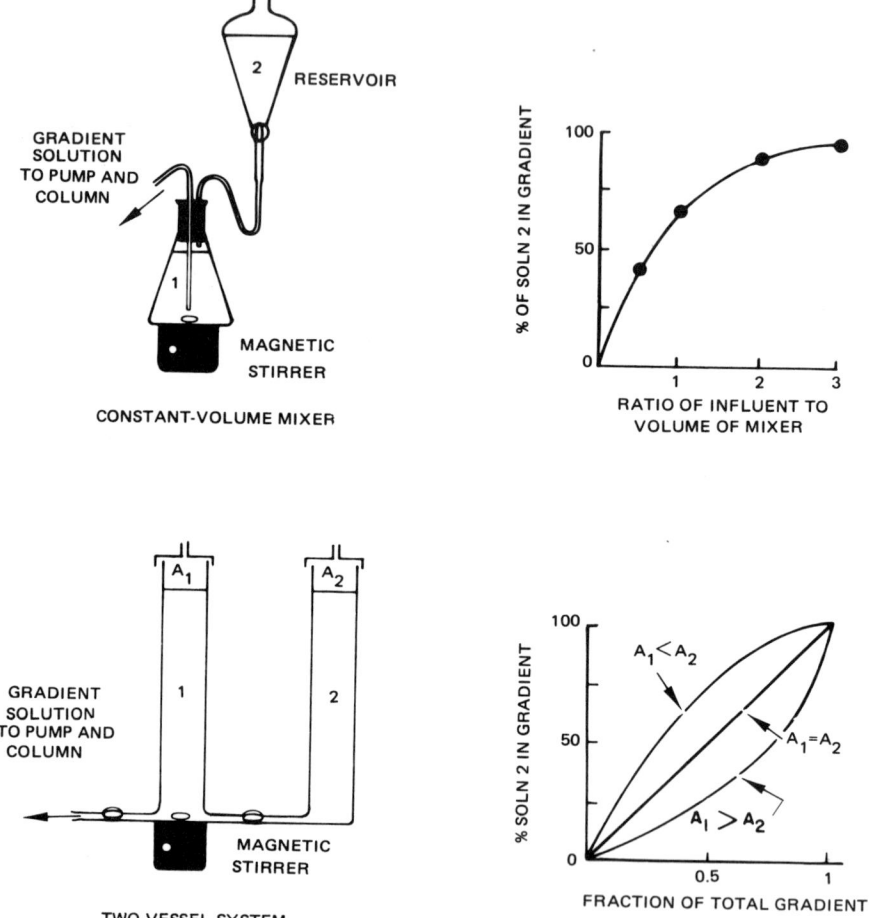

Figure 4.1. Apparatus for the production of gradients and nature of the gradient from each type of apparatus.

Certain aspects of this type of gradient should be noted. It is simple to set up and use. It is versatile because the solution in the reservoir may be changed repeatedly with resulting change in the gradient. The ability to alter the gradient at will is particularly valuable because the gradient changes only slowly after a volume equal to twice the volume of the mixer has been withdrawn.

This happens because the gradient is convex with a greater initial slope that constantly decreases (Figure 4.1). Unless the chromatogram can be completed with the initial limiting solution in the reservoir, it should be changed.

The slope of the gradient is dependent on two factors: (1) the differential in concentration between solutions in mixer and reservoir, and (2) the volume of the mixer. If the differential between the solutions is slight, the slope of the gradient will be slight, and vice versa. Similarly, if the volume of the mixer is changed but the differential between solutions is the same, the smaller the mixer, the greater the slope.

These comments about the slope of the gradient are applicable only to concentration gradients, for example, of NaCl. They do not apply exactly to pH gradients. Although pH gradients are easily produced, their exact form will depend on the range of the gradient in relation to the pK of the buffer. Thus, a pH gradient which passes through the pH of the pK will have a different form than one further removed on the titration curve.

It should be obvious that the actual gradient through the chromatogram is dependent on the size of the chromatographic column, the size of the gradient vessel, and the differential in composition of the solutions.

2. Multivessel Systems

Bock and Ling (1954) describe experimental setups for many types of gradient. One of the most useful involves two parallel cylinders that are connected at the bottom and one of which (the mixer) has an outlet to the column (Figure 4.1). If both cylinders are filled to equal height (hydrostatic equilibrium) with different solutions and liquid is withdrawn, the level will fall in both cylinders, but the solution from the second cylinder (the reservoir) must pass through the mixer. Consequently, if the mixer is stirred, a gradient is formed.

This type of gradient is self-limiting. If the chromatogram is incomplete when the gradient has been used, continuation with

some type of solution or a second gradient is necessary. Constant-
volume gradients are always convex, but the relationship of the
cross-sectional areas of the two vessels of a two-vessel system
will determine whether the gradient is convex, linear, or concave:
if the area of the mixer is the lesser, the gradient is convex; if
the areas are equal, the gradient is linear; if the area of the
mixer is the greater, the gradient is concave (Figure 4.1). The
gradient from a two-vessel system may be calculated from the equation

$$C = C_2 - (C_2 - C_1)(1 - x)^{A_2/A_1}$$

where C, C_1, and C_2 have the same meanings as above, x is the fraction of the *total* gradient used, and A_1 and A_2 are the cross-
sectional areas of the vessels (Bock and Ling, 1954).

The comments in the preceding section about slope of gradient
are also essentially applicable to the two-vessel system.

Peterson and Sober (1959) describe a multivessel system with
nine connecting cylinders and furnish formulas for calculating the
resulting gradient, but such complex devices usually are unnecessary in hemoglobin chromatography.

C. Flow Rate of Developer

In most analytical procedures of hemoglobin chromatography, columns
of 1 cm diameter with lengths of 20-60 cm are used. For the most
part, flow rates have been limited on columns of this diameter to
5-20 ml hr^{-1} [6-25 ml hr^{-1} $(cm^2)^{-1}$ of cross-sectional area]. As
a consequence, when the total effluent through the column may be
500-1000 ml or more, the duration of the chromatogram may be several days. Recent work suggests that such flow rates are more
conservative than necessary. For example, CM-cellulose chromatography (Schroeder *et al.*, 1976) was done at 50 ml hr^{-1} routinely,
and even 75 ml hr^{-1} [60-90 ml hr^{-1} $(cm^2)^{-1}$ of cross-sectional area]
without worsening the separations; consequently, the chromatogram
was completed in an overnight period. In microchromatographic
procedures for hemoglobin to be described in Chapters 7, 8, and 9,

flow rates are also on the order of 50 ml hr^{-1} $(cm^2)^{-1}$ of cross-sectional area.

Flow rates through chromatographic columns may be controlled by means of the hydrostatic head of the developing solvent. However, more regular rates (and, therefore, more regular fraction size with a time-controlled fraction collector) result if a constant-volume pump or a peristaltic pump is used. Such pumps are offered commercially by many firms.

I. INTRODUCTION

Boardman and Partridge (1953, 1955) were the first to separate hemoglobins by ion-exchange chromatography. By means of Amberlite IRC-50, they separated adult bovine carbonmonoxyhemoglobin from fetal sheep carbonmonoxyhemoglobin, adult from fetal sheep carbon-monoxyhemoglobin, and bovine carbonmonoxyhemoglobin from bovine ferrihemoglobin. Perhaps more important than the actual separations that they achieved was their careful examination of such factors as pH, ion concentration, and temperature in relation to chromatographic behavior. They set forth many of the principles that have been discussed in Chapter 3.

Prins and Huisman (1955) and Huisman and Prins (1955, 1957) elaborated on Boardman and Partridge's method and applied it to the separation of human hemoglobin variants. Their "cuvette method," which employed a rectangular column that was 0.5 X 3 cm in cross-sectional dimension and 15 cm long, permitted the separation of hemoglobins F, A, S, and C.

Morrison and Cook (1955) then showed a separation of normal adult human hemoglobin into three components. A further description of their studies is detailed in a subsequent paper (Morrison and Cook, 1957).

Allen, Schroeder, and Balog (1958) modified chromatographic procedures for hemoglobin on Amberlite IRC-50 and especially examined the heterogeneity that had been observed by Morrison and Cook (1955) and Prins and Huisman (1956). Allen and coworkers used phosphate instead of citrate or citrate-phosphate buffers as developers and introduced the addition of KCN to the developers and spectrophotometric determination at the Soret band at 415 nm for increased sensitivity in detection of minor components.

Prins (1959) gives an excellent summary of the state of the art in hemoglobin chromatography after the first few years of its application.

Although Amberlite IRC-50 was the ion exchanger of choice in these early studies, CM-cellulose was introduced in extensive experiments by Huisman and collaborators (Huisman and Meyering,

1960; Huisman, Martis, and Dozy, 1958; Meyering *et al.*, 1960) and
later the use of CM-Sephadex in hemoglobin chromatography was described by Dozy and Huisman (1969).

Although the carboxyl group is common to Amberlite IRC-50, CM-
cellulose, and CM-Sephadex, each ion exchanger is individualistic
in its ability to separate the components of a mixture of hemoglobins. This chapter will describe in detail what appears to be the
best set of conditions that has been devised in the course of time
for hemoglobin chromatography on each of the three and will provide
references additional to these brief historical comments.

If the reader wishes to know whether or not any of these procedures has been applied to a specific hemoglobin variant, the
Appendix should be consulted.

II.   CHROMATOGRAPHY ON AMBERLITE
      IRC-50

This description of procedure is a combination of the procedure
of Allen, Schroeder, and Balog (1958) with modifications and am-
plifications that resulted from the use of the method by Clegg
and Schroeder (1959), Jones and Schroeder (1963a,b), and Schnek
and Schroeder (1961).

A.   Procedure

1. *Developers*

In the procedures that are being described, development usually
has been by means of a single developer that is chosen to provide
the characteristics necessary to bring about the separation of the
mixture under consideration. The six developers whose compositions
and pH values are given in Table 5.1 have been used. The developing power of the solutions is determined by the pH and ionic concentration; the higher the pH and ionic concentration, the more
rapidly will a hemoglobin pass down the column. Developers can be
and have been tailored to meet specific needs. For example, Developer 6 was designed to retard the movement and bring about the

Table 5.1 Composition of Buffers Used as Developers in Amberlite
IRC-50 Chromatography

No.	pH (at 25°C)	Concn. Na$^+$ (M)	Concn. KCN (M)	Buffer (g liter^{-1})		
				NaH$_2$PO$_4 \cdot$H$_2$O	Na$_2$HPO$_4$	KCN
1	7.22 ± 0.02	0.075	0.01	3.45	3.55	0.65
2	7.18 ± 0.02	0.0625	0.01	3.45	2.66	0.65
3	7.02 ± 0.02	0.050	0.01	3.45	1.77	0.65
4	6.91 ± 0.02	0.050	0.01	4.14	1.42	0.65
5	6.85 ± 0.05	0.055	0.01	4.14	1.77	0.65
6	6.66 ± 0.04	0.050	0.01	4.60	1.17	0.65

separation of hemoglobins such as Hb H and Hb Bart's that are
electrophoretically fast moving in alkaline medium. It would be
useful with similarly moving hemoglobins such as Hb J and Hb N.
On the other hand, Developer 1 or even one of higher pH or ionic
concentration is necessary for Hb C. Specific examples of the
use of various developers will be presented below in Section II.B
of this chapter.

2. Preparation of Ion Exchanger

At the time of the experiments of Allen, Schroeder, and Balog
(1958), commercially available Amberlite IRC-50 contained a wide
range of particle sizes. Therefore, it was first purified by the
method of Hirs, Moore, and Stein (1953) and sifted wet in hydrogen form to obtain material that passed through a 200-, but was
retained by a 250-mesh sieve. After washing with 3 M HCl and
water, the resin is ready for further preparation. Jones and
Schroeder (1963a) report no apparent difference in chromatographic
behavior when the particles were of mesh size 200-250 or 250-325,
and Schroeder and Holmquist (1966) used mesh size 200-325 with
success. At the present time, resin of limited mesh range is
available in the sodium form from such suppliers as Bio-Rad Laboratories (Richmond, CA 94804).

In preparation for pouring the column, the resin is suspended in about 5 times its volume of the appropriate developer. With continuous mechanical stirring, the pH is now titrated to that of the developer. If the resin has been prepared by sifting and is in hydrogen form, this titration is done with 40% NaOH. If the commercially sized resin in sodium form is used, it need only be suspended in buffer and titrated with phosphoric acid. Stirring is continued for some time and additional titration is made if the pH changes. The resin is then washed thoroughly on a funnel and suspended in a volume of buffer equal to three times the volume of settled resin. The resin is allowed to settle and any fine particles ("fines") are removed by decantation. If necessary, the settling is repeated until all fines are gone. The final ratio of settled resin to supernatant buffer should be 1:3.

3. *Pouring and Equilibration of the Column*

As described, this procedure requires that the chromatography be done at 6°C. Consequently, if the column is not poured in a jacketed tube and maintained at this temperature with a constant temperature bath, the column should be poured in a cold room.

For analytical purposes, columns have commonly been 1 X 35 cm. After a column has been poured as described in Chapter 2, Section V, it is equilibrated by passing through the appropriate developer at a flow rate of about 10 ml $(cm^2)^{-1}$ of cross-sectional area until at least 1 liter $(cm^2)^{-1}$ of cross-sectional area has been used.

4. *Development of the Chromatogram*

After a column has been poured and equilibrated, the top 1-2 cm is carefully stirred with an even, circular motion and allowed to re-settle before the sample is applied.

The sample is prepared from hemolysate from which cell debris has been removed and that has been dialyzed in the cold against 100 or more times its volume of the chosen developer. For a 1 X 35 cm column, 30-60 mg hemoglobin in 0.5-2.0 ml solvent is convenient.

After the top of the column has been stirred and resettled into a flat, even surface, all solvent except a 2-3 mm layer above the column is removed. The sample is then carefully layered onto the column and developer itself above the hemoglobin solution. When flow is started, the hemoglobin solution passes easily and evenly into the column. If the resin bed should be disturbed during addition of the sample, skewed zones that are likely to result can usually be corrected by stirring the sample carefully into the upper centimeter or two and settling it before development is begun.

Flow rate of developer through 1 X 35 cm columns has generally been started at 5-6 ml hr^{-1} and then doubled at some time during the chromatogram. This is one aspect of this system that has not been examined in detail. It is probable that significantly faster flow rates would not greatly worsen the separations (see Chapter 8).

Fraction sizes of 1 or 2 ml have normally been collected.

The chromatographic behavior of hemoglobins on Amberlite IRC-50 is greatly dependent upon the temperature. Consequently, in some procedures the temperature has been increased to speed the elution of certain components. When this has been done, the chromatogram has been stopped, and water of the desired temperature has been circulated through the jacket for 30 min before the flow is again started.

5. Reequilibration and Reuse

Chromatograms may be run repeatedly on Amberlite IRC-50 columns without the necessity of repouring. If there has been no change in temperature in the preceding chromatogram and if the developer with which the column is equilibrated is to be used again, the next chromatogram may be started as soon as the top has been stirred and resettled.

If the temperature has been changed and/or if another developer is to be used, the temperature must be adjusted and at least 1 liter of developer passed through a 1 X 35 cm column in order to reequilibrate. After this, the column may be used again.

B. Examples of Chromatography on
    Amberlite IRC-50

When Developer 1 is used with normal adult human hemoglobin, there
is rapid movement through the column but, nevertheless, a separation into two components (Figure 5.1). The great sensitivity of
the method to the developing conditions is evident from Figure 5.2
in which Developer 2 was substituted. Hb A_I emerges almost as
rapidly but Hb A_{II} is retarded and well separated from Hb A_I. If
hemoglobin from the cord blood of a normal newborn infant is chromatographed under the same conditions as in Figure 5.2, the major
peak now is at the position of Hb A_I and the smaller component is

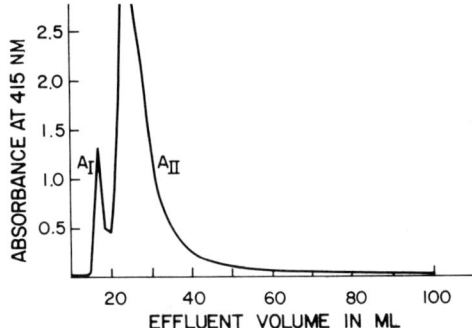

Figure 5.1. Chromatogram of normal adult human hemoglobin on a
1 X 35 cm column of Amberlite IRC-50 with Developer 1. (Adapted
from Allen, Schroeder, and Balog, 1958.)

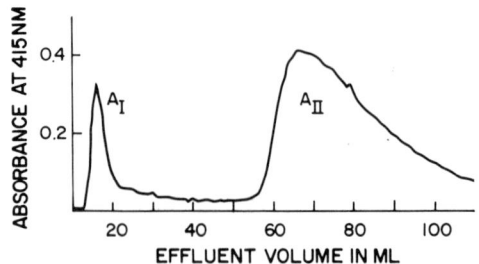

Figure 5.2. Chromatogram of normal adult human hemoglobin on a
1 X 35 cm column of Amberlite IRC-50 with Developer 2. (Adapted
from Allen, Schroeder, and Balog, 1958.)

now Hb A_{II}. The major component from cord blood at the position of Hb A_I in Figure 5.2 is fetal hemoglobin or Hb F and minor, related components; the Hb A_{II} from the adult and infant are identical.

The heterogeneity of Hb A_I and Hb F becomes apparent when these are rechromatographed with Developer 4 in order to retard their movement still further (Figures 5.3 and 5.4). As may be seen

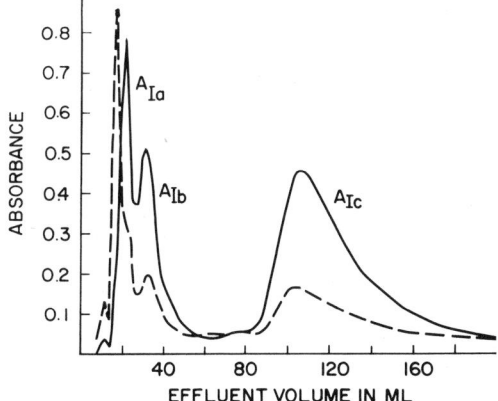

Figure 5.3. Rechromatography of zone A_I on a 1 X 35 cm column of Amberlite IRC-50 with Developer 4. Solid line = 415 nm; broken line = 280 nm. (Adapted from Allen, Schroeder, and Balog, 1958.)

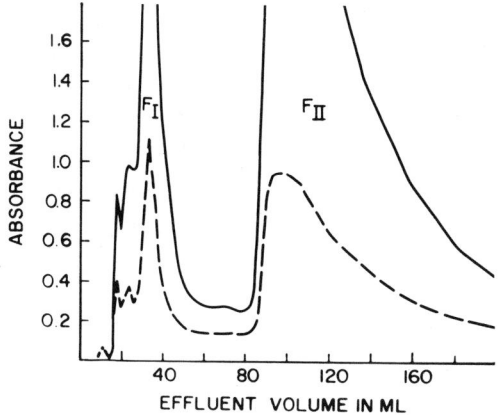

Figure 5.4. Rechromatography of zone F on a 1 X 35 cm column of Amberlite IRC-50 with Developer 4. Solid line = 415 nm; broken line = 280 nm. (Adapted from Allen, Schroeder, and Balog, 1958.)

Hb A_{Ib} and Hb F_I as well as Hb A_{Ic} and Hb F_{II} have virtually identical chromatographic behavior with Developer 4.

Acetylation (or possibly glycosylation) at the N-terminus of
the γ chains of Hb F_I (Schroeder *et al.*, 1962; Stegink, Meyer, and
Brummel, 1971) apparently is the sole difference from Hb F (Hb F =
Hb F_{II} = Hb F_o) or $\alpha_2\gamma_2$ and is responsible for the charge difference, and, hence, the chromatographic separation. Hb A_{Ic} has glucose
in Schiff's base linkage to the N-termini of the β chains of Hb A
(Hb A = Hb A_{II} = Hb A_o) or $\alpha_2\beta_2$ (Bookchin and Gallop, 1968; Bunn
et al., 1975; Holmquist and Schroeder, 1966; Koenig, Blobstein,
and Cerami, 1977). This topic is discussed further in Chapter 8.
Under any conditions of chromatography on Amberlite IRC-50, Hb F
and Hb A_{Ic} are inseparable. Thus, the 0.5-1.0% of Hb F that is
present in the hemoglobin of the normal adult will cochromatograph
with Hb A_{Ic}. A different chromatographic system is required in
order to separate these two components (Chapters 6 and 9). McDonald
et al. (1978) have separated Hb A_{Ia} into two components which are
glycosylated and contain phosphate. According to Krishnamoorthy,
Gacon, and Labie (1977) as well as McDonald *et al.* (1978), Hb A_{Ib}
contains no phosphate.

When Developer 5, which produces a somewhat more rapid movement of components than does Developer 4, is applied to normal adult
hemoglobin, the chromatogram of Figure 5.5 results. Hb A_{Ia} and Hb
A_{Ib} do not separate from each other but do separate from Hb A_{Ic}.
Hb F cochromatographs with Hb A_{Ic}, and if the sample should contain an elevated percentage of Hb F, this will manifest itself by
an Hb A_{Ic} percentage above the normal 4-6%. Hb A_{Id} and Hb A_{Ie} may
be artifacts. If the chromatogram had been continued at 6°C, the
Hb A_{II} would have emerged as a very spread-out zone in a large volume of effluent. However, because the temperature was raised to
28°C when the leading edge of Hb A_{II} neared the bottom of the column, Hb A_{II} emerged rapidly and separated from Hb A_{III}. Hb A_{III}
contains Hb A_2 $(\alpha_2\delta_2)$, but this procedure is inaccurate for quantitation; the level of Hb A_2 normally is only 2-3%. It is important
that Hb A_{II} approach the bottom of the column before the temperature

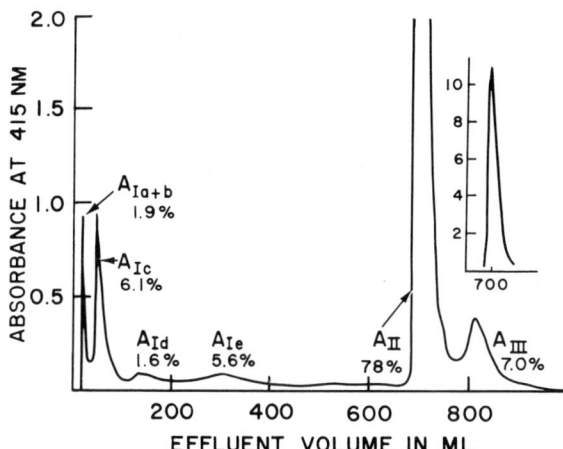

Figure 5.5. Chromatogram of whole adult hemoglobin on a 1 X 35 cm column of Amberlite IRC-50 with Developer 5. The inset shows that portion of the chromatogram near 700 ml effluent volume on a reduced vertical scale. The chromatogram was warmed from 5-6°C to 28°C after 675 ml effluent. (From Clegg and Schroeder, 1959, with permission.)

is raised. If this is not done, small amounts of hemoglobin on the apparently white column below the Hb A_{II} will elute together quickly to form a sharp extraneous peak immediately ahead of Hb A_{II}.

Other examples of IRC chromatography may be seen in the sources already cited.

C. Comments

The conditions of chromatography may be varied to produce excellent separations. The separation of Hb F and Hb A is complete. The capacity of the columns is excellent and these may be reused almost indefinitely.

Among disadvantages of the method is the slow equilibration of the column which requires 1 liter of developer for a 1 X 35 cm column when developer or temperature is changed. Zones on this sorbant are generally nonsymmetrical and tail considerably as Figures 5.1 and 5.5 show. Jones and Schroeder (1963a) studied the effect of loading on the movement of the zones. Thus, when Hb S was chromato-

graphed on a 1 X 35 cm column with Developer 1, the rate of movement of the zone was constant when the load was varied from 25 to
100 mg. Below this range, the zones moved more slowly and above it
more rapidly. This behavior is not a serious drawback because of
the rather wide range of load with constant rate of movement, but
cannot be entirely ignored if two hemoglobins are being compared.
A more serious problem that could lead to misinterpretation is that
of "double zoning." Figure 5.6 suggests a heterogeneity of the main
component of Hb S from a patient with sickle cell anemia. Despite
extensive study, the cause of the phenomenon was not definitely delineated. However, it was concluded that both zones were of identical primary structure that underwent reversible change to two
chromatographic forms.

     Chromatography on Amberlite IRC-50 generally has been done near
5°C, probably because Boardman and Partridge (1955) and Huisman and
Prins (1955) obtained better recoveries at low temperatures than at
room temperature. Inasmuch as their experiments were done without

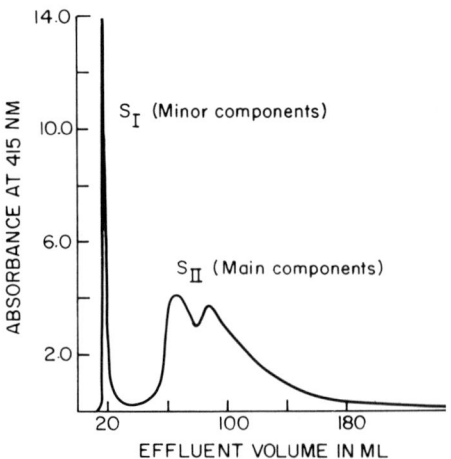

Figure 5.6. Double zoning of hemoglobin S_{II}. A 50-mg sample of
hemoglobin from a patient with sickle cell anemia was chromatographed on a 1 X 35 cm column of Amberlite IRC-50 at 6°C with Developer 1. A slowly moving minor component, S_{III}, is not shown.
(From Jones and Schroeder, 1963a, with permission.)

the addition of KCN to the developers, their poorer recoveries at
room temperature may have resulted from partial conversion to ferrihemoglobin during chromatography. Any ferrihemoglobin would become
much more tightly fixed and would not be removed unless large volumes of developer were used. However, Trivelli, Ranney, and Lai
(1971) have used Developer 6 at room temperature instead of Developer 5 at 6-7°C with success in the separation of Hb A_{Ia+b} and Hb
A_{Ic} from Hb A. By judicious modification of conditions, Amberlite
IRC-50 chromatography can be done at room temperature. Because
chromatographic behavior on Amberlite IRC-50 is greatly influenced
by temperature and because increased temperature has been used to
remove hemoglobins that are strongly fixed (see Figure 5.5), chromatography at room temperature requires a lowered pH and/or ionic
concentration.

In applications where the presence of KCN might be undesirable
for succeeding experiments, carbonmonoxyhemoglobin, the more stable
derivative, probably provides an adequate alternative form for
chromatography.

D. Other Procedures of Amberlite
    IRC-50 Chromatography

Although the work of Boardman and Partridge (1955) and of Morrison
and Cook (1955) was important in initiating chromatographic procedures for hemoglobins, other investigators do not seem to have
used their exact methods.

Huisman and Prins (1955, 1957) devised methods with similarity
to those of Boardman and Partridge. They describe a column method
but emphasize a cuvette method, in which the solid phase is rectangular in a tube that is 0.5 X 3 cm in cross-section and 15 cm
long. Citrate buffers of adjusted sodium ion concentration were
used; the pH was 6.5 for the column method and 6.0 for the cuvette
method.

Huisman and Meyering (1960) made an extensive comparison of
Amberlite IRC-50 and CM-cellulose chromatography of hemoglobin.
They have used the only gradient system for hemoglobin chromatography

on Amberlite IRC-50. Their column, almost twice the length of that
of Allen, Schroeder, and Balog (1958), was equilibrated with Developer 4, which was also present in a small constant-volume mixer
(50 ml) to produce a steep gradient to Developer 4 at twice the
sodium ion concentration. The zones were more symmetrical by this
method.

Itano and Robinson (1960) modified the methods of Allen and
co-workers (1958) by the use of a low pH, a higher sodium ion concentration, and stepwise development with increase in sodium ion
concentration.

Hill *et al.* (1962) also modified the methods of Allen, Schroeder,
and Balog (1958) to use a lower pH (6.42), higher sodium ion concentration, no KCN, and carbonmonoxyhemoglobin. Their modifications
have been applied by Konigsberg and Lehmann (1965) and Ranney *et al.*
(1968) for the isolations of Hb M by stepwise change to 0.5 M sodium
ion concentration.

In a somewhat similar manner, Hayashi *et al.* (1968) substituted
potassium phosphate buffers at pH 7 apparently without KCN, a procedure that has been used by Wajcman, Belkhodja, and Labie (1972)
for the isolation of Hb Setif and by Wajcman, Labie, and Schapira
(1973) for the isolation of Hb Tours. Stamatoyannopoulos *et al.*
(1973) report the isolation of Hb Olympia on Amberlite IRC-50 with
0.1 M potassium phosphate at pH 7 without further elaboration of
conditions.

The increased interest in the quantitation of Hb A_{Ia+b} and
Hb A_{Ic} in diabetics prompted Trivelli, Ranney, and Lai (1971) to
modify Allen, Schroeder, and Balog's procedure for use at room temperature. Separation was made with Developer 6 at room temperature
at a flow approximately 10 times that of Allen and co-workers.
Fitzgibbons, Koler, and Jones (1976) have made small alterations
in the method of Trivelli, Ranney, and Lai in quantitative examination of Hb A_{Ia+b} and Hb A_{Ic} in relation to red-cell age in normal
and diabetic subjects. Jones, Koler, and Jones (1978) describe a
microchromatographic procedure for the determination of Hb A_{Ic} on

Amberlite IRC-50 at room temperature. Chapter 8 discusses these
modifications in detail. Bunn *et al.* (1975) isolated Hb A_{Ic} for
chemical studies by Trivelli, Ranney, and Lai's method.

III. CHROMATOGRAPHY ON CM-SEPHADEX

The description of this procedure follows that of Dozy and Huisman
(1969). The method is useful for the separation and quantitation
of many major and minor hemoglobins with nearly identical electrophoretic properties. The time required to complete a chromatogram
is a disadvantage which limits its use as a routine procedure.

A. Procedure

1. Developers

These are 0.05 M in Tris and have pH values between 6.5 and 8.0.
A 1.0 M Tris stock solution is prepared by dissolving 121.1 g Tris
in 1000 ml distilled water; the pH of this solution is not adjusted.
The solution is often slightly yellow and filtration over activated
charcoal will remove the color. Developers are prepared from this
stock solution by proper dilution, 100 mg KCN is added per 1000 ml,
and the pH is adjusted with 1 M maleic acid. The developers are
identified by their pH values, for instance, Tris-maleic acid, pH
7.0.

2. Preparation of Ion Exchanger

The material is CM-Sephadex (C-50, capacity: 4.5 ± 0.5 meq g^{-1})
manufactured by Pharmacia Fine Chemicals, Inc. (501 Fifth Ave.,
New York, NY).

This cation exchanger is suspended in a large volume of the
first developer (Tris-maleic acid, pH 6.5), and the suspension is
stirred repeatedly for several hours. The supernatant is removed
and replaced by the same developer. This washing procedure is repeated twice. The suspension of one part settled ion exchanger and
one part supernatant is stored at room temperature, and the supernatant is replaced with fresh developer once weekly.

3. *Pouring and Equilibration of the Column*

The CM-Sephadex suspension is poured in sections to form a 0.9 X 60 cm column. The column is equilibrated overnight at 4°C with the Tris-maleic acid, pH 6.5 developer at the rate of 30 ml hr^{-1}.

4. *Development of the Chromatogram*

This part of the procedure is conducted at 4°C, preferably in a cold room. After hemolysate has been prepared from a freshly collected blood sample, an appropriate volume which contains 70-150 mg hemoglobin is dialyzed overnight against 500 ml Tris-maleic acid, pH 6.5 at 4°C. This sample is applied on top of the CM-Sephadex and gently stirred into the top 1-cm portion of the column. Any hemoglobin on the inside wall of the glass tube is carefully washed down with 0.5-1.0 ml Tris-maleic acid, pH 6.5. The same developer is layered over the resin and the glass tube is filled with the same solution. The column is connected with a 250-ml constant-volume mixing flask which is filled with Developer Tris-maleic acid, pH 6.5. A pH gradient is applied by supplying Developer Tris-maleic acid, pH 7.0 to the mixer. The flow rate is maintained at 18-20 ml hr^{-1} by means of a pump and 6-ml fractions are collected. It is not advisable to increase the flow rate because of the undesirable packing of the CM-Sephadex under increased pressure. The selection of additional Tris-maleic acid developers with higher pH values, such as 7.3, 7.5, 7.7, and 8.0, and their time of introduction depends on the elution pattern observed and the sample to be chromatographed. The time to complete the chromatogram of a sample which contains, among others, the electrophoretically slow-moving Hb C is nearly 7 days. Recovery of the hemoglobin applied to the column is better than 99.5%.

5. *Reequilibration and Reuse of the Cation Exchanger*

Because CM-Sephadex as used in these chromatographic experiments is difficult to reequilibrate without changes in its chromatographic

properties, it is discarded and new cation exchanger is used for
each chromatogram.

B. Examples of Chromatography on
    CM-Sephadex

Figure 5.7 illustrates the separation of the components in an artificial mixture of hemoglobins A, F, S, C, and A_2. On the basis of
electrophoretic examination, each component was pure, except for
the A_2 and S fractions, which were contaminated with each other to
a minor extent. The elution pattern is similar to but slightly different from that observed in CM-cellulose chromatography (Section
IV of this chapter). The separation of hemoglobins F and A is comparable to that observed in Amberlite IRC-50 chromatography.

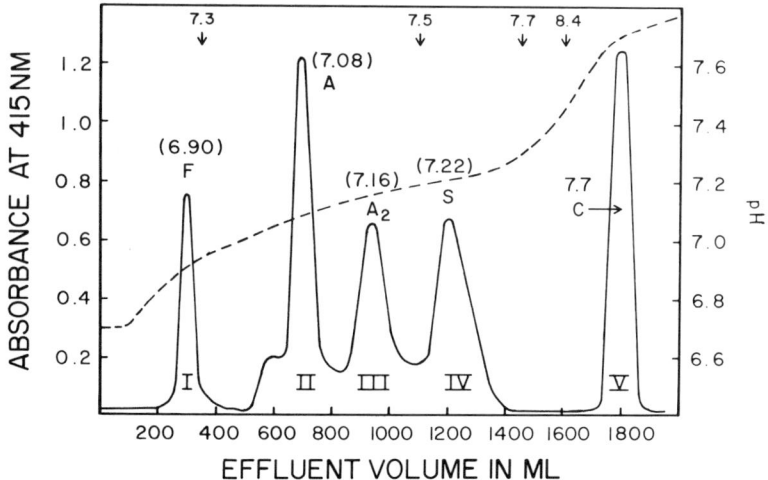

Figure 5.7. Chromatographic separation of the hemoglobins F, A, A_2,
S, and C in an artificial mixture on a column of CM-Sephadex at 4°C.
After an initial pH gradient with 0.05 M Tris-maleic acid developers
of pH 6.7-7.0 was applied, the second buffer of this gradient system
was replaced by additional developers with increasing pH values (7.3,
7.5, 7.7, 8.4) as indicated. The values in parentheses are the pH
elution values of the individual zones. The broken line indicates
the pH of the effluent. The flow rate was 20 ml hr^{-1}. (From Dozy
and Huisman, 1969, with permission.)

The separation of Hb F from a cord-blood hemolysate and that
of the hemoglobins in samples from patients with certain hemoglobinopathies are shown in Figure 5.8; the same samples were analyzed
by DEAE-Sephadex chromatography. From the examination of these
chromatograms, the following observations may be made: (1) Hb F
is observed as two incompletely separated components which are eluted
at pH values of 6.91 and 6.94, respectively; (2) the total amount of
Hb F as calculated from a CM-Sephadex chromatogram corresponds rather
well with that from a DEAE-Sephadex chromatogram; (3) because the
separation of the Hb A_2, Hb S, and Hb C was (almost) complete, Hb A_2
can be quantitatively determined in patients who are heterozygous
or homozygous for Hb C; and (4) the resolution by CM-Sephadex chromatography is greater than that by DEAE-Sephadex chromatography. This
is particularly apparent in the chromatograms of the β-thalassemia
heterozygotes (Cases W.O. and G.S.).

C.   Other Procedures of CM-Sephadex
     Chromatography

Abraham *et al.* (1975) have slightly modified the above system to
study the hemoglobins of patients with sickle cell anemia. The
chromatogram is developed at room temperature with a slightly decreased flow rate and a specific series of developers. About 60 mg
hemoglobin, which has been dialyzed against the pH 6.5 developer,
is applied to the 0.9 X 60 cm column. The pH gradient is formed by
introducing Developer Tris-maleic acid, pH 7.0 into the 250-ml
mixing flask which is filled with Developer Tris-maleic acid, pH
6.5. After 24 hr, Developer Tris-maleic acid, pH 7.0 is replaced
by one with pH 7.3. This procedure is repeated on two consecutive
days with developers of pH 7.5 and 7.8, respectively. A flow rate
of 12 ml hr^{-1} is maintained. The time required to complete the
chromatogram is 4-5 days.

Figure 5.9 illustrates two chromatograms of the hemoglobin from
two SS patients with different quantities of Hb F. The separation
of the five hemoglobin zones is complete, and quantitation is readily possible.

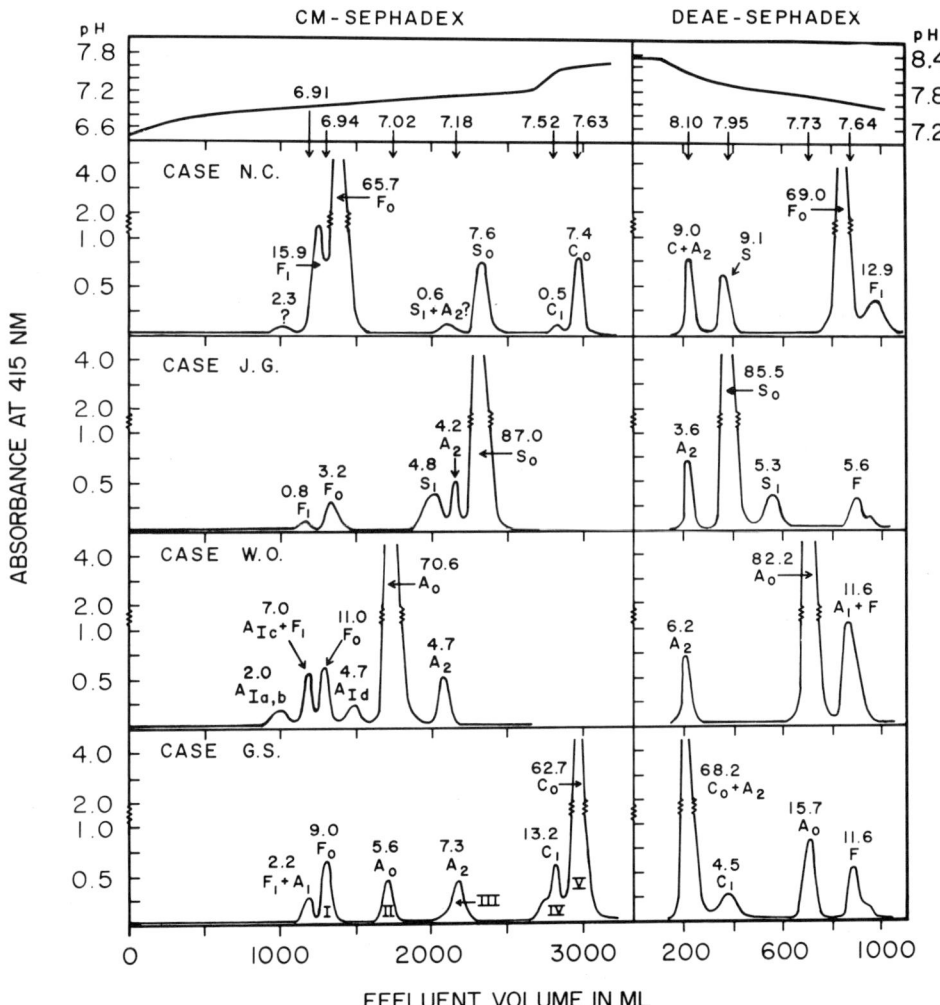

Figure 5.8. Chromatographic separation of the hemoglobins in samples that contain Hb F on columns of CM-Sephadex and DEAE-Sephadex. Case N.C., cord blood from a baby heterozygous for the hemoglobins S and C; Case J.G., sickle cell anemia; Case W.O., β-thalassemia trait; and Case G.S., Hb-C-β-thalassemia trait. The values are the percentages of the individual hemoglobins; the pH of elution of each zone is given at the top of the first chromatogram. (From Dozy and Huisman, 1969, with permission.)

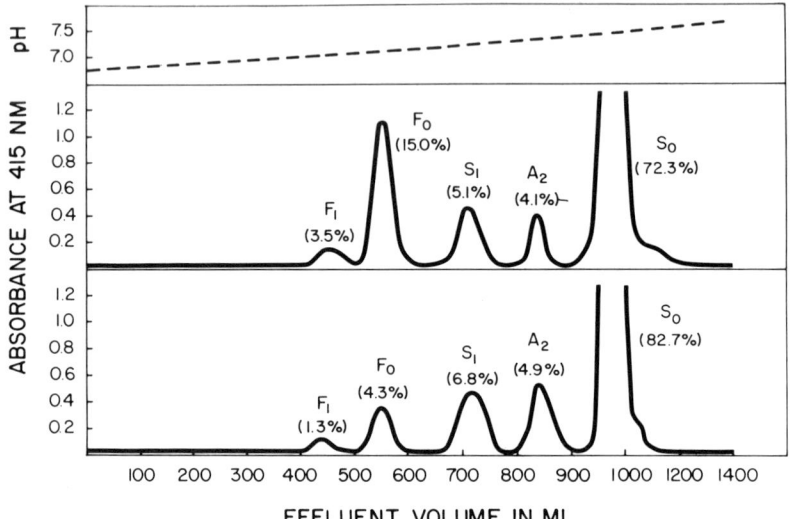

Figure 5.9. Separation of hemoglobins from two patients with sickle cell anemia by modified CM-Sephadex chromatography. For details, see text.

IV. CHROMATOGRAPHY ON CM-CELLULOSE WITH PHOSPHATE DEVELOPERS

The original procedure was developed during 1958-1960 by Huisman, Meyering, and collaborators (Huisman and Meyering, 1960; Huisman, Martis, and Dozy, 1958; Meyering *et al.*, 1960). These early experiments often resulted in partial separations that were greatly improved when the original CM-cellulose was replaced by the microgranular, preswollen type. The procedure, to be described here, follows the outline of Huisman (1972b) which was further detailed by Huisman and Wrightstone (1974).

A. Procedure

1. *Developers*

These are 0.01 M sodium phosphate solutions with 100 mg KCN liter^{-1} and pH values of 6.7, 6.9, 7.1, 7.4, 7.6, 7.8, and 8.2. These developers are prepared by mixing 1 liter 0.01 M NaH_2PO_4 (1.56 g

$NaH_2PO_4 \cdot 2H_2O$) and 100 mg KCN liter^{-1} (the KCN should be dissolved
in the NaH_2PO_4 solution just prior to the preparation of the developers) with an appropriate volume of 0.01 M Na_2HPO_4 (3.582 g Na_2HPO_4
$\cdot 12H_2O$) and 100 mg KCN liter^{-1} until the desired pH value is obtained. The seven developers are stored in closed bottles and are
identified by their pH value.

2. Preparation of the Ion Exchanger

The material is CM-cellulose (CM-52: microgranular, preswollen),
which is manufactured by Whatman Biochemicals and marketed in the
United States by Whatman (Clifton, NJ).

About 100 g CM-52 is mixed with 500 ml Developer pH 6.7. The
suspension is left overnight whereafter the supernatant is discarded. After this procedure has been repeated twice (except that
equilibration is limited each time to a few hours), the cation exchanger is stored at room temperature in the same developer as a
suspension of 2 parts settled material and 1 part supernatant
solution.

3. Pouring and Equilibration of the Column

A 1.8 X 35 cm column is poured from the suspension in sections and
is equilibrated with Developer pH 6.7 for a few hours at 30 ml hr^{-1}.

4. Development of the Chromatogram

After hemolysate has been prepared from a freshly collected blood
sample, an appropriate volume which contains 50-60 mg hemoglobin
is dialyzed overnight at 4°C against Developer pH 6.7. This sample is applied and gently stirred into the top 5 mm of the column.
Hemoglobin which clings to the inside wall of the glass tube is
carefully washed down with 0.5 ml Developer pH 6.7. This developer is layered over the resin, and the glass tube is filled with
the same solution. The column is connected with a 250-ml constant-
volume mixing flask which contains Developer pH 6.9. Development
is at room temperature with a pH gradient that is formed by introducing Developer pH 7.4 (or Developer pH 7.1 if a slower pH gradient

is desired) into the mixer. The flow rate of the column is maintained at 16-20 ml hr^{-1} with a pump, and fractions are collected at
20-min intervals. After 24-30 hr, when the total elution volume is
at least 400 ml, the developer in the reservoir is replaced by one
with a higher pH value (usually pH 7.6). The elution of a few hemoglobin variants (for example, Hb C) requires the use of developers
with a pH of 7.8 or even 8.2.

5. *Reequilibration and Reuse of*
    the Cation Exchanger

It is advisable to accumulate used CM-52 from several columns, thoroughly clean it with Developer pH 8.2, wash it repeatedly with distilled water, and finally reequilibrate it with Developer pH 6.7.

B. Examples of Chromatography on
    CM-Cellulose with Phosphate Developers

Figure 5.10 illustrates the chromatograms of hemoglobins from four
patients who are heterozygous for (a) Hb C ($\alpha_2\beta_2$ 6 Glu → Lys), (b)
Hb E ($\alpha_2\beta_2$ 26 Glu → Lys), (c) Hb O-Arab ($\alpha_2\beta_2$ 121 Glu → Lys), or
(d) Hb Agenogi ($\alpha_2\beta_2$ 90 Glu → Lys), respectively. The patient with
Hb C also produced considerable quantities of Hb F. Hemoglobins A
and F, and the minor A_1 and F_1 components are eluted in the first
400 ml effluent and are poorly separated. Hb A_2, which is eluted
next, and Hb E have the same chromatographic properties. However,
hemoglobins Agenogi, O-Arab, C, and their respective minor hemoglobins, which are eluted at specific volumes and pH values, are well
separated from each other and from Hb A_2. Therefore, it is possible
to quantitate Hb A_2 in Hb C-containing samples.

Figure 5.11 illustrates separation of the hemoglobins from newborns who were heterozygous (1) for Hb S ($\alpha_2\beta_2$ 6 Glu → Val), (2)
for Hb G-Philadelphia (α_2 68 Asn → Lys β_2), (3) for Hb F-Port Royal
($\alpha_2\gamma_2$ 125 Glu → Ala), or (4) for Hb F-Malta I ($\alpha_2\gamma_2$ 117 His → Arg).
The separation of the abnormal hemoglobins from the normally occur-
ring hemoglobins F and A is complete in all four cases. However,

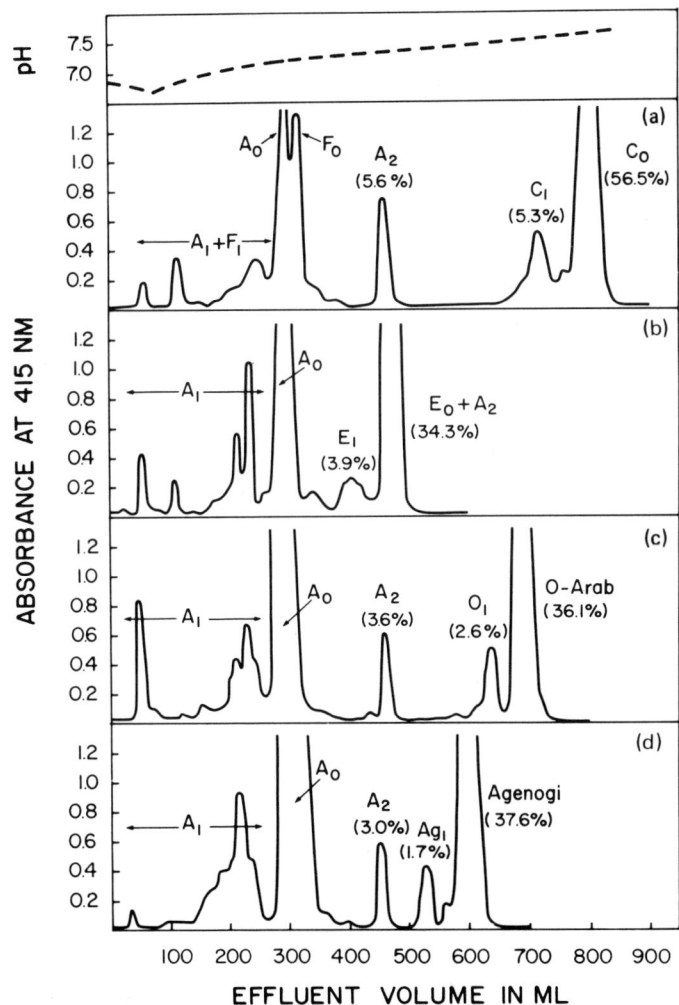

Figure 5.10. Chromatographic separation on columns of CM-cellulose of the hemoglobins in hemolysates which contain various slow-moving variants. For details, see text.

the incomplete separation of Hb F and Hb A makes it impossible to quantitate the fetal hemoglobins by this method or to distinguish between babies with sickle cell trait (AS) and sickle cell anemia.

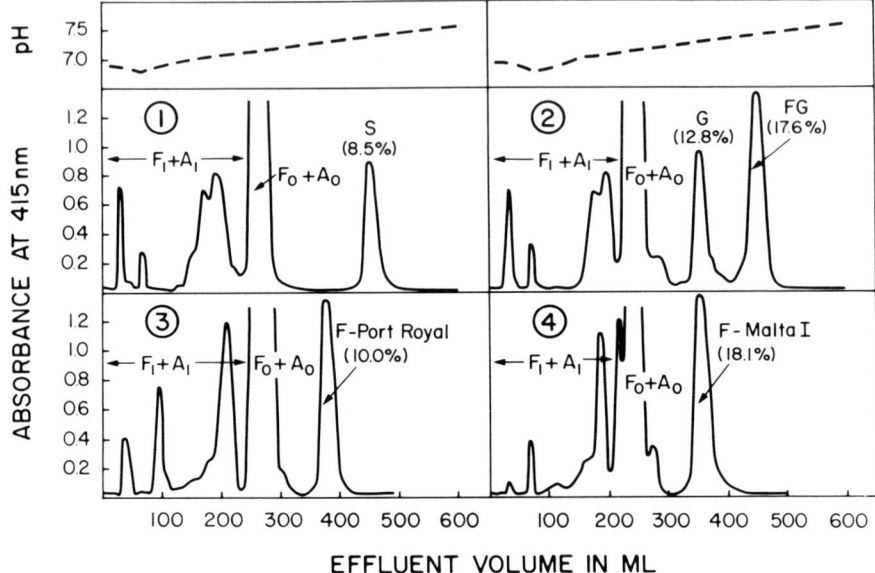

Figure 5.11. Chromatographic separation on columns of CM-cellulose of the hemoglobins in hemolysates from newborns with different hemoglobin variants. For details, see text.

C. Comments

The results that have been depicted illustrate how hemoglobin variants with the same substitution at different positions may have significantly different chromatographic properties. At pH 6.5, the protein is attached to the functional groups of the CM-cellulose mainly through its α-NH_3^+ and ϵ-NH_3^+ groups. Consequently, a change in pH towards the isoelectric point of the hemoglobin causes it to be desorbed. Because the replacement, for instance, of a glutamyl residue by a lysyl residue results in an increase in the positive net charge at pH 6.5, the variant is more strongly adsorbed by the cation exchanger and a higher pH is required for its desorption. Theoretically, the replacement of a glutamyl residue by a lysyl residue in different positions of either the α or β chains would influence the adsorption-desorption properties of the variants in an equal manner, unless secondary changes alter the net charge of

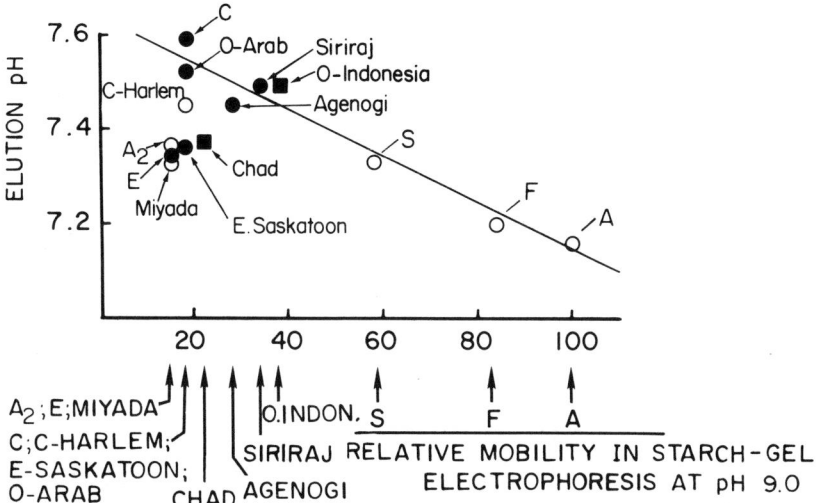

Figure 5.12. Relationship between pH value at elution and mobility in starch-gel electrophoresis at pH 9. Closed squares and closed circles refer to α- and β-chain variants with a Glu → Lys substitution, respectively. (From Huisman and Wrightstone, 1974, with permission.)

the protein and its binding to the ion exchanger. This binding is not only dependent on the amount of charge on the protein at a given pH and fixed ionic strength of the solvent, but also on the accessibility of the functional groups on the surface of the protein. If it is assumed that the hemoglobins are not subject to major configurational changes at the pH values of the chromatographic experiments (6.7-7.8), the difference in chromatographic affinity of variants with supposedly similar isoelectric points (see Figure 5.10) presumably is caused by differences in accessibility of certain charged groups.

Figure 5.12 illustrates the relation between the chromatographic and electrophoretic properties of several α and β chain variants (many with a Glu → Lys substitution) by comparing the pH values at elution with their relative electrophoretic mobilities. A direct relationship is evident for many variants (including Hb F) but some, such as Hb E ($\alpha_2\beta_2$ 26 Glu → Lys), Hb E-Saskatoon ($\alpha_2\beta_2$ 22

Glu → Lys), and Hb Chad (α_2 23 Glu → Lys β_2), join Hb A_2 and Hb Miyada in having a considerably lower pH value at elution than is expected. These characteristic properties can be most useful when a chromatographic method for the separation of certain hemoglobin variants has to be selected.

V. CHROMATOGRAPHY ON CM-CELLULOSE
    WITH BIS-TRIS AND SODIUM
    CHLORIDE DEVELOPERS

As is apparent from the preceding section, CM-cellulose provides many excellent separations. In contrast to other cation exchangers, however, the separation of Hb F and Hb A is minimal. For the most part, phosphate has been the buffer of choice and development has been by pH gradient. When CM-cellulose is used with bis-Tris to control pH and an NaCl gradient to elute the hemoglobins, the separation of Hb F and Hb A is much improved and other separations are excellent as well. The use of CM-cellulose under such conditions is described by Schroeder, Pace, and Huisman (1976).*

A. Procedure

1. *Developers*

Development in this procedure uses a linear gradient of NaCl at a constant pH and bis-Tris concentration. Table 5.2 provides data on the compositions of the solutions. These solutions are of general usefulness for qualitative and quantitative assessment of a given sample of hemoglobin which has normal and/or common abnormal hemo-globins. Thus, Developer BT-1 is used for preparation of the ion exchanger and equilibration of the column as well as the developer in the mixer of the gradient. The developer in the other vessel of the gradient is selected on the basis of the composition of the hemo-globin. If Hb C ($\alpha_2\beta_2$ 6 Glu → Lys) or Hb O-Arab ($\alpha_2\beta_2$ 121 Glu → Lys)

---

*Much of this section is paraphrase or direct quotation from Schroeder, Pace, and Huisman (1976) by permission.

Table 5.2 Composition of Buffers for Use in Developers in CM-Cellulose Chromatography with bis-Tris[a]

No.	Concn. bis-Tris (M)	Concn. NaCl (M)	Concn. KCN (%)	Grams liter^{-1}		
				bis-Tris	NaCl	KCN
BT-1	0.03	0.03	0.01	6.28	1.75	0.1
BT-2	0.03	0.085	0.01	6.28	4.97	0.1
BT-3	0.03	0.12	0.01	6.28	7.00	0.1

[a]All solutions are adjusted to pH 6.1 with 6 M HCl.

is expected to be present, Developer BT-3 is chosen, but if these are known to be absent, then Developer BT-2 may be used with a lesser total volume of gradient to give the same gradient of NaCl concentration.

2. Preparation of Ion Exchanger

The CM-cellulose for this procedure is the microgranular, preswollen material designated as CM-52 and supplied by Whatman (Clifton, NJ).

A weighed amount of CM-52 is suspended in 6 times weight per volume of the desired developer and allowed to settle. The supernatant solution with any fines is decanted and the procedure is repeated to remove any remaining fines. After resuspending, the pH is adjusted if necessary after settling and resuspension. The final slurry for pouring the column is adjusted so that the ratio of settled ion exchanger to supernatant solution is 1:2.

3. Pouring and Equilibration of the Column

All operations for this procedure are done at room temperature. After a 1 X 20 cm column has been poured, the column is equilibrated with 50-100 ml of the chosen developer in the course of 1-2 hr at 50 ml hr^{-1} with a pump. Back pressure in these columns is slight so that a peristaltic pump is adequate.

4. Development of the Chromatogram

The sample is prepared by dialyzing hemolysate overnight at 4°C
against a large volume of the selected developer. It is then carefully layered onto the column and developer is layered above the
hemoglobin solution.

The plumbing between column and gradient vessel is filled with
developer which is present in the mixer of the gradient device. If
Hb C is or may be present and the full gradient is to be used, then
650 ml Developer BT-1 is placed in the mixer and 650 ml Developer
BT-3 in the second vessel. However, the shortened gradient uses
375 ml Developer BT-1 in the mixer and 375 ml Developer BT-2 in the
second vessel. Nevertheless, the slope of the gradient is identical in both cases.

When less common hemoglobins are to be separated, the conditions may be altered to suit the properties of the components.
Variants with an electrophoretic mobility greater at alkaline pH
than that of Hb A may be retarded in their passage through the column by decrease of pH, bis-Tris concentration, and slope of NaCl
gradient which may begin at zero concentration. The reverse changes
will speed the movement of electrophoretically slow-moving components. Examples will be given below.

The flow rate of developer is 50 ml hr^{-1} and 5-ml fractions of
the effluent are collected. Conductance, or osmolality, is used to
determine the NaCl concentration at intervals of 10 fractions.

5. Reequilibration and Reuse

The columns may be reequilibrated with 100 ml of the developer of
choice and used at least 5 times. It may be desirable to replace
the top centimeter of the column after several chromatograms.

B. Examples of Chromatography on CM-cellulose
    with bis-Tris-NaCl Developers

The chromatograms that result from the hemoglobin mixtures in a
variety of normal and abnormal hematological conditions are depicted in Figures 5.13 to 5.15.

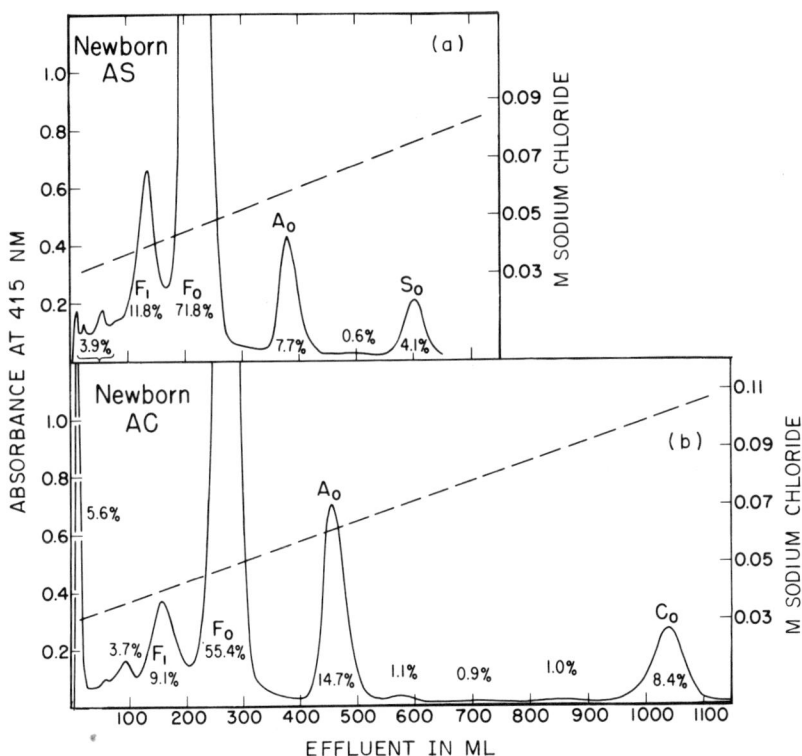

Figure 5.13. Separation of hemoglobins in cord blood from an infant with sickle cell trait (a) and an infant with Hb C trait (b) on CM-cellulose. Shortened and full gradients, respectively, were used. (From Schroeder, Pace, and Huisman, 1976, with permission.)

The data in Figure 5.13a derive from the cord blood of a newborn infant with sickle cell trait. The separation of Hb F, Hb A, and Hb S is excellent in this system and Hb F_1 separates from Hb F_0 as it does in most chromatographic systems. The newborn child with Hb C trait has the pattern in Figure 5.13b. The use of both the full and shortened gradients is shown in this figure.

Hemoglobin from a normal adult yields the chromatogram of Figure 5.14a. In addition to Hb A_2, a number of other minor components separate from the major Hb A_0 of the normal individual. The very rapidly moving zones may contain some enzymes of the red cell

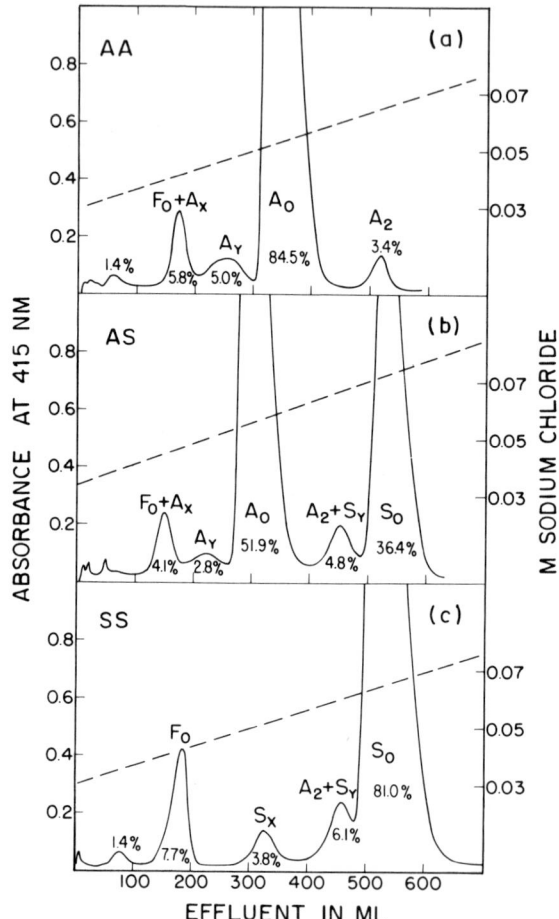

Figure 5.14.   Separation of hemoglobins of a normal adult (a) and
of adults with sickle cell trait (b) and sickle cell anemia (c) by
the shortened gradient on CM-cellulose.   (From Schroeder, Pace, and
Huisman, 1976, with permission.)

(Huisman and Meyering, 1960) as well as the pyridoxal complex

(Srivastava, van Loon, and Beutler, 1972).   Traces of Hb F_o and

minor components of Hb A are not separable.   Because no detailed

study has been made of minor components and because they (with

the exception of A_2 and F_1) have not been correlated with minor

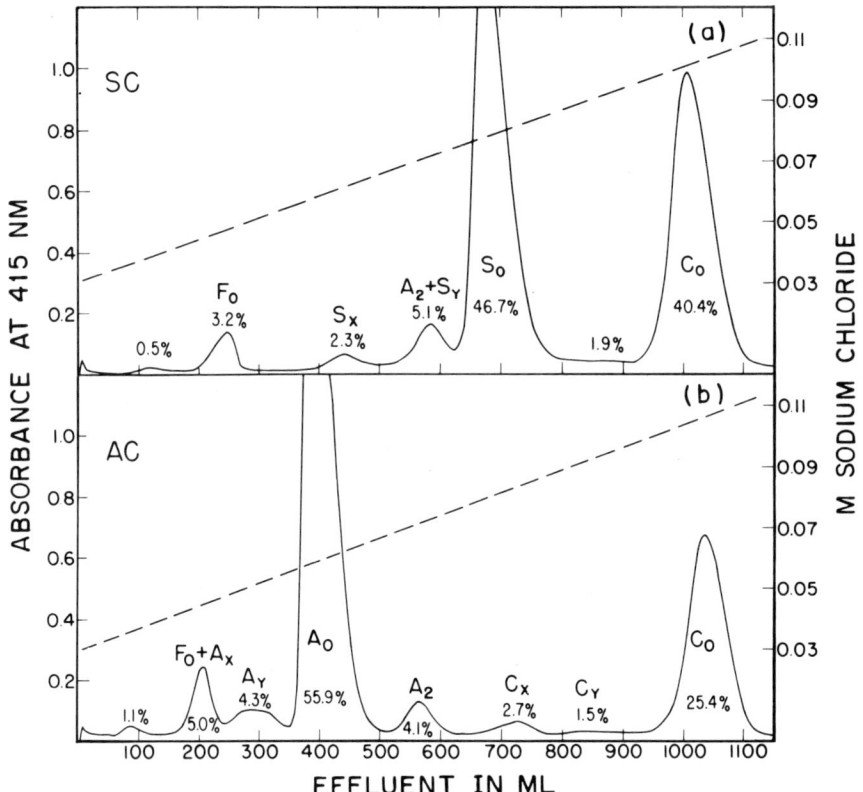

Figure 5.15. Separation of hemoglobins of adults with SC disease (a) and Hb C trait (b) by the full gradient on CM-cellulose. (From Schroeder, Pace, and Huisman, 1976, with permission.)

components as detailed by other chromatographic methods or by electrophoresis, they are labeled as A_x, A_y, etc., in the figures.

When Hb S is present, the separations in Figures 5.14b (sickle cell trait) and 5.14c (sickle cell anemia) are obtained. A minor component(s) related to Hb S (S_y) overlaps Hb A_2, although sometimes there may be partial separation. Hb A_2, therefore, cannot be quantitatively determined in the presence of Hb S by this method.

When Hb C is present and the full gradient is used, Hb C emerges, as shown in Figures 5.13b, 5.15a, and 5.15b, before the end of the gradient. The separation of Hb A_2 from Hb C is excellent and permits the quantitative determination of Hb A_2 in the presence of Hb C. From a comparison of various chromatograms in Figures 5.14 and 5.15, it is apparent that Hb A_o in an AS sample will be contaminated with a minor component of Hb S (note the position of S_x in Figures 5.14c and 5.15a, as compared to Hb A_o in Figure 5.14b). On starch-gel electrophoresis at pH 9.0, S_x does not behave like A_o but is heterogeneous and moves like S_o and F.

In other hematological conditions, the nature of the chromatogram will, of course, be altered. Thus, in the adult heterozygote for the hereditary persistence of fetal hemoglobin (HPFH) who will have 15-30% Hb F, the increased Hb F of such a heterozygote will become apparent through a greater peak of $F_o + A_x$ in Figure 5.14a. Because of the variability of Hb F in sickle cell anemia, the distinction between S-HPFH and sickle cell anemia cannot be made on the basis of chromatographic criteria alone. If Hb S coexists with β-thalassemia, Hb A is absent in the type termed $S\text{-}\beta^0$-thal and present in $S\text{-}\beta^+$-thal. The chromatogram of the former would have much the appearance of Figure 5.14c but because A_2 and S_y do not separate, a distinction between SS and $S\text{-}\beta^0$-thal on the basis of the percentage of Hb A_2 would not be possible. However, in $S\text{-}\beta^+$- thal, the S_x peak would contain Hb A_o and be present to the extent of 15-25%, and Hb F would vary from case to case.

When other hemoglobins were chromatographed (not depicted), an Hb D (possibly D-Los Angeles or $\alpha_2\beta_2$ 121 Glu $\rightarrow$ Gln) moved more rapidly than Hb S in a distinct peak, although incompletely separated from Hb S. Hb Lepore (probably Lepore-Washington) has the mobility of Hb A_2 as does Hb E. Hb O-Arab is distinguishable from Hb C.

C. Comments

Bis-Tris has been chosen as a buffer for these procedures because its pK_a is 6.5 and therefore the buffer capacity at the pH of choice is good. Because bis-Tris is a relatively expensive chemical, it would be advantageous to reduce its concentration in the developers.

Satisfactory chromatograms result if the concentration of bis-Tris
is reduced from 0.03 to 0.01 M. However, the reproducibility of
the point of emergence of a given hemoglobin seems to be somewhat
more variable probably because of poorer pH control at the lower
concentration of bis-Tris.

The conditions of chromatography have been chosen to provide
a relatively rapid movement of hemoglobins on the column. If an
electrophoretically fast-moving hemoglobin at alkaline pH were
present, it would be virtually unadsorbed under these conditions.
However, a pattern such as that in Figure 5.13a or other figures
can be translated along the volume axis by changing the NaCl gradient at constant pH and bis-Tris and KCN molarity, and thus the
rate of movement of more rapidly moving components can be retarded.
Two examples are given in Figures 5.16a and b which present data
from a Thailander in whom Hb H (β_4) was detected by starch-gel
electrophoresis, and from his newborn child in whose hemoglobin Hb
Bart's (γ_4) was present. Because of the presence of these electrophoretically fast-moving hemoglobins, the gradients were modified.
For the chromatogram in Figure 5.16a, the NaCl concentration in the
two buffers was 0.01 M and 0.08 M in 0.03 M bis-Tris and 0.01% KCN
at pH 6.1, respectively; the total gradient was 1000 ml; and
the column had been equilibrated with the former. For that in Figure 5.16b, the column was equilibrated with 0.01 M bis-Tris, 0.0
NaCl, and 0.01% KCN; 25 ml of this solution was used for initial
development with a subsequent 1000-ml gradient between 0.0 NaCl and
0.08 M NaCl in 0.03 M bis-Tris and 0.01% KCN at pH 6.1. The movement of hemoglobins is retarded under these conditions but the NaCl
molarity at which a given peak emerges is within the range that is
observed with the usual gradient. It is interesting to compare the
chromatogram of Figure 5.16a with that of Figure 5.14a. In both
chromatograms, the slope of the gradient was identical but the start
at 0.01 M NaCl in the chromatogram of Figure 5.16a instead of 0.03
M NaCl retarded the elution of Hb A_o about 275 ml effluent. Similar
retardation occurs if the gradient is started at 0.03 M NaCl but the

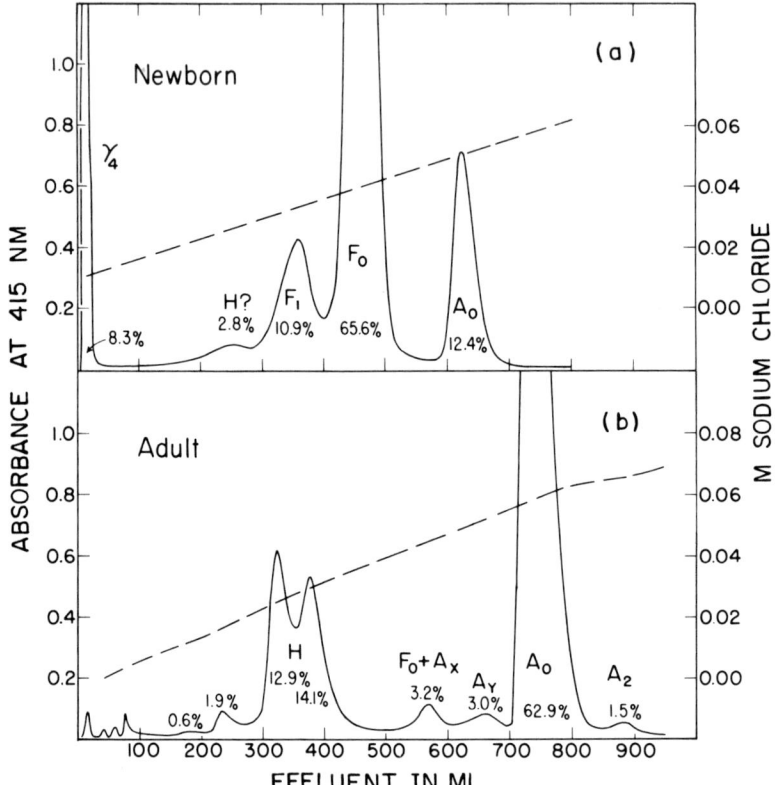

Figure 5.16. Separation of hemoglobins in a newborn infant with Hb Bart's (a) and in the father who has Hb H disease (b) with a modified gradient on CM-cellulose (see text). (From Schroeder, Pace, and Huisman, 1976, with permission.)

bis-Tris is 0.01 M instead of 0.03 M. In summary, an increase in pH, NaCl, or bis-Tris concentration, or in slope of the NaCl gradient speeds the movement of any hemoglobin and vice versa.

A slower flow rate or a longer column does not improve separations. Most of the chromatograms in the figures used a flow rate of 25 ml hr^{-1}, but equally good separations result when the flow rate is 50 ml hr^{-1}. Consequently, the chromatogram is complete in about 1 day. In fact, the flow rate may be increased to 75 ml hr^{-1} without significant deterioration in the separation. Visual observations of the movement of the hemoglobins on the column suggest that most of the separation occurs in the upper 10 cm, and that each hemoglobin washes virtually unretarded through the lower half of the column.

I.   INTRODUCTION

Peterson, Sober, and associates (Peterson and Sober, 1956; Sober
and Peterson, 1958; Sober *et al.*, 1956) and Fahey and collaborators
(Fahey and Goodman, 1960; Fahey, McCoy, and Goulian, 1958) were
among the first to introduce weak anion exchangers, such as diethylaminoethyl (DEAE) cellulose, triethylaminoethyl (TEAE) cellulose,
and ECTEOLA-cellulose as chromatographic media for the separation
of proteins. Usually buffer gradients of increasing concentration

and/or decreasing pH resulted in the separation of many proteins.
When serum proteins were chromatographed, they appeared in the effluent in order of increasing electrophoretic mobility at alkaline pH.

Similar anion exchangers have been used for the separation of
many normal and abnormal hemoglobins both of human and animal origin. Most useful are the two diethylaminoethyl anion exchangers
DEAE-Sephadex and DEAE-cellulose, whereas TEAE-cellulose and ECTEOLA-
cellulose have been much more limited in their application in this
field. The commercially available anion exchangers have improved
markedly during the years; DEAE-cellulose was originally available
as Selectacel-DEAE, Type 40^{+}, and varied in properties from lot to
lot; now a preswollen microgranular material which is known as Whatman DE-52 gives improved and reproducible separations. Likewise,
DEAE-Sephadex was initially available as an A-50 medium, poorly sized
preparation, but now has been developed into a beaded form of particle size 40-120 μ which gives more constant flow rates and reproducible chromatograms. In this chapter, we will describe in detail the
application of these anion exchangers to the separation of the normal hemoglobins A, F, and A$_2$, and of many of their variants.

Reference is again made to the Appendix as a source of information about the application of any of these procedures to a specific
hemoglobin variant.

II. CHROMATOGRAPHY ON DEAE-SEPHADEX

The method that will be described was developed initially by Huisman
and Dozy (1965) and improved by Dozy, Kleihauer, and Huisman (1968).
The beaded form of the DEAE-Sephadex whose use was introduced in
1968 resulted in a marked improvement in the resolution of various
hemoglobins.

A. Procedure

1. *Developers*

These are 0.5 M in Tris and have pH values between 6.5 and 8.5. A
1.0 M Tris stock solution is prepared by dissolving 121.1 g Tris in

approx. 0.8 liter distilled water; the pH is adjusted to 8.5 with 4 M HCl, and the volume to 1 liter. If the solution is light yel- low as is often the case, filtration over activated charcoal will remove the color. Developers are prepared from this stock solution by proper dilution, 100 mg KCN liter^{-1} is added, and the pH adjusted with 2 M HCl. Each developer is best identified by its pH value.

2. Preparation of Ion Exchanger

The material originally used was DEAE-Sephadex, A-50 medium (Phar- macia Fine Chemicals, Inc., 501 Fifth Avenue, New York, N.Y.) but it was soon replaced by a beaded form of this anion exchanger: DEAE-Sephadex A-50, capacity 3.5 ± 0.5 meq g^{-1}, particle size 40- 120 μ. Numerous lots of this material have been tested with iden- tical results.

This dry material is suspended in a large volume of Tris-HCl buffer, pH 8.5, and the suspension is stirred repeatedly for several hours. The pH of the suspension is adjusted with stirring to 8.3- 8.5 with 1.0 M Tris solution. The supernatant solution is removed and replaced by the Tris-HCl, pH 8.5 developer. This washing pro- cedure is repeated twice whereafter the suspension is stored at room temperature in the same developer. The ratio of settled ion exchanger to supernatant solution should be about 1:1.

3. Pouring and Equilibration of
the Column

A 0.9 X 50-55 cm column is poured in sections. Although equilibra- tion of the column is not necessary, it has certain advantages for the separation of hemoglobin variants with high isoelectric points. When equilibration is done, Developer pH 8.5 is passed through for 24 hr at a flow rate of 15-20 ml hr^{-1} at room temperature.

4. Development of the Chromatogram

After hemolysate has been prepared from a freshly collected blood sample, an appropriate volume which contains 30-50 mg hemoglobin is dialyzed against 500 ml Tris-HCL, pH 8.5 overnight at 4°C. This sample is applied on top of the DEAE-Sephadex, and gently stirred into the top 1-cm portion of the exchanger. The hemoglobin on the inside wall of the glass tube is carefully washed down with about 0.5 ml Tris-HCl, pH 8.5. The same developer is layered over the resin and the tube is filled with the same solution. A continuously decreasing pH gradient is obtained by introducing a Tris-HCl developer of lower pH into a 500-ml constant-volume mixer.

The starting developer in the mixer is Tris-HCl, pH 8.3, and the pH of the influent developer is 7.9, but may vary from 8.1 to 7.0 depending upon the hemoglobin to be chromatographed. About 24 hr later, the first influent buffer is replaced by one with a pH that is 0.2-0.3 pH units lower; when necessary, this procedure is repeated again 24 hr later. Thus, the changes are made ad lib on the basis of experience alone. A flow rate of 15-20 ml is maintained with a pump; fractions are collected at 20-min intervals. The chromatogram is developed at room temperature. The recovery of the hemoglobin from the column approaches 100%, provided that hemolysate from a freshly collected blood sample is chromatographed.

5. *Reequilibration and Reuse*

Reequilibration of the DEAE-Sephadex in the column, although possible, has not been very successful, mainly because of changes in the flow rate, and is therefore not recommended. Better results are obtained when used resin, from several columns, is washed on a Buchner funnel with distilled water, 1 M NaCl solution (1-2 liters will be sufficient for material in a funnel with a diameter of 25-35 cm with material to a depth of about 4 cm), again with distilled water (2-4 liters), and finally with Tris-HCl, pH 8.5 (about 2 liters). The DEAE-Sephadex will be slightly yellow but will have (nearly) the same chromatographic properties as new material. New DEAE-Sephadex should always be used in analytical chromatograms.

B. Examples of Chromatography on DEAE-Sephadex

Figures 6.1 and 6.2 give several examples of chromatographic separation that can be obtained with this procedure. The chromatograms

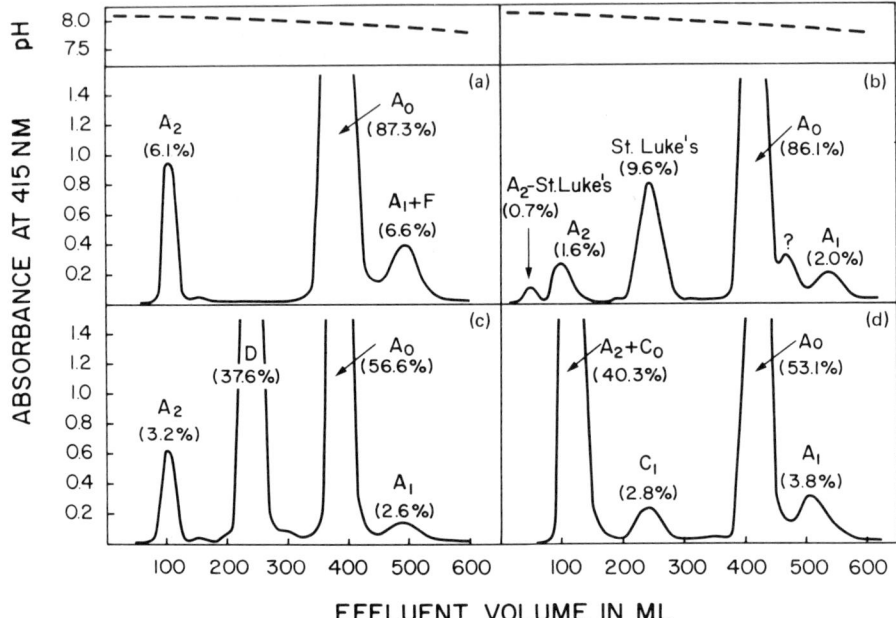

Figure 6.1. Chromatographic separation of the hemoglobins of a β-thalassemia heterozygote (a), an Hb St. Luke's heterozygote (b), an Hb D heterozygote (c), and an Hb C heterozygote (d). For details, see text.

of Figure 6.1 illustrate several points: (a) the person with β-thalassemia trait has an elevated level of Hb A_2; (b) the Hb St. Luke's heterozygote has an α-chain variant (α_2 95 Pro → Arg β_2), and therefore, there are two minor hemoglobins which contain δ chains; (c) the person with an Hb D-Los Angeles ($\alpha_2\beta_2$ 121 Glu → Gln) heterozygosity has a chromatogram indistinguishable from that of sickle cell trait; and (d) the Hb C ($\alpha_2\beta_2$ 6 Glu → Lys) of a C trait does not separate from Hb A_2. Separation of the hemoglobin zones is often complete except for Hb F and the minor Hb A_1. The hemoglobins are eluted in rather broad zones and their elution is entirely pH dependent (for instance, the Hb St. Luke's zone is eluted in 70-80 ml effluent at an average pH value of 8.0).

Figure 6.2 depicts the separation of fast-moving hemoglobins from Hb A. The variants that were studied are: Hb Hope ($\alpha_2\beta_2$ 136

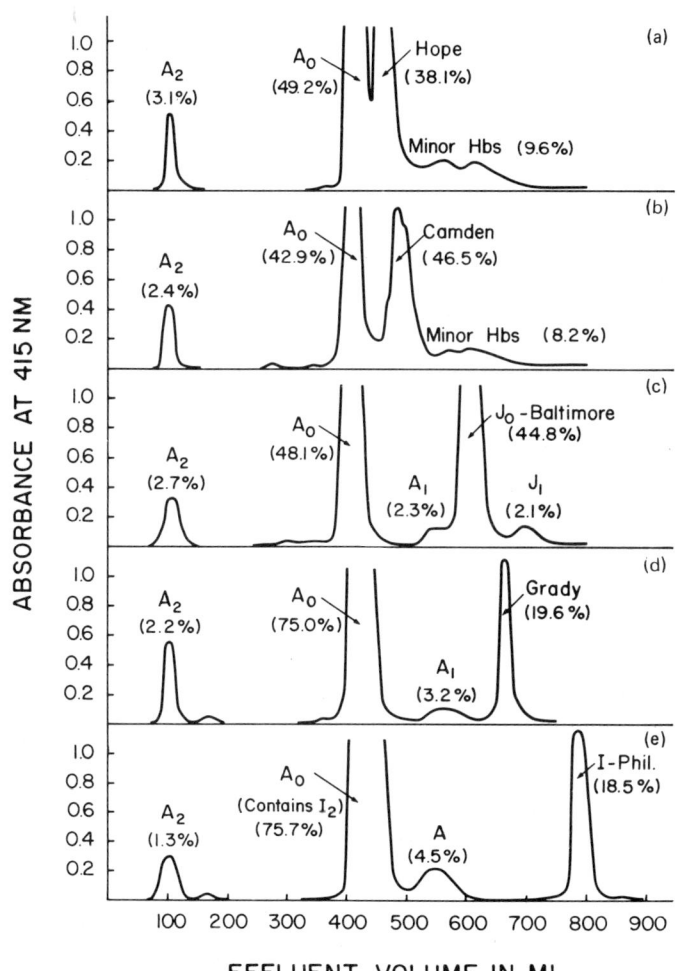

Figure 6.2. Chromatographic separation of the hemoglobins of persons heterozygous for any one of five different fast-moving hemoglobin variants. For details, see text.

Gly → Asp) (6.2a); Hb Camden ($\alpha_2\beta_2$ 131 Gln → Glu) (6.2b); Hb J-Baltimore ($\alpha_2\beta_2$ 16 Gly → Asp) (6.2c); Hb Grady ($\alpha_2\beta_2$ with α chains being extended through an insertion of a Glu-Phe-Thr segment between residues 118 and 119) (6.2d); and Hb I-Philadelphia (α_2 16 Lys → Glu β_2) (6.2e). The separation of Hb A and Hb Hope and of Hb A and Hb Camden, although adequate for many purposes, is incomplete.

The same is the case with Hb J-Baltimore and the minor Hb A_1. The
minor hemoglobins A_2-Grady ($\alpha_2^{\text{Grady}} \delta_2$) and A_2-I-Philadelphia
($\alpha_2^{\text{I}} \delta_2$) cochromatographed with Hb A. Again, the zones are rather
broad and elution of the hemoglobins is pH dependent (pH values are
not included in Figure 6.2; often when a method becomes routine in
a laboratory, certain measurements that are not critical for the
completion of the chromatogram are inadvertently omitted). Of interest is the difference in the chromatographic behavior of the
hemoglobins Hope and J-Baltimore; both variants have a Gly → Asp
substitution but at different positions in the β chain.

C. Comments

During the past 10-15 years, the DEAE-Sephadex procedure probably
has been the most popular chromatographic method for quantitating
abnormal hemoglobins and for isolation of certain variants on a
large scale.

        Dozy, Kleihauer, and Huisman (1968) compared the results of
several chromatograms with those obtained by the starch-gel electrophoretic procedure. The elution of a certain hemoglobin component in this type of chromatography appears to be determined exclusively by the pH of the developing gradient. The accuracy of
the measurement of the effluent pH depends on many factors, such
as temperature, the pH meter, operator, and others. It is therefore not surprising that relatively large ranges of effluent pH
values were observed when several chromatograms were compared.
Such a comparison is presented in Figure 6.3. The solid curves
summarize data of chromatograms which were selected at random [the
numbers of analyses were 21 B_2 ($\alpha_2 \delta_2$ 16 Gly → Arg); 82 A_2; 42 S_0;
4 S_1; 79 A_0; and 63 A_1], whereas, the hatched curves are composed
from data of chromatograms which were developed under more consis-
tent conditions, such as constant temperature, one operator, and
one type of pH meter. In most instances, the separation of the
various hemoglobin fractions was complete while the differences in
the pH values of elution were constant.

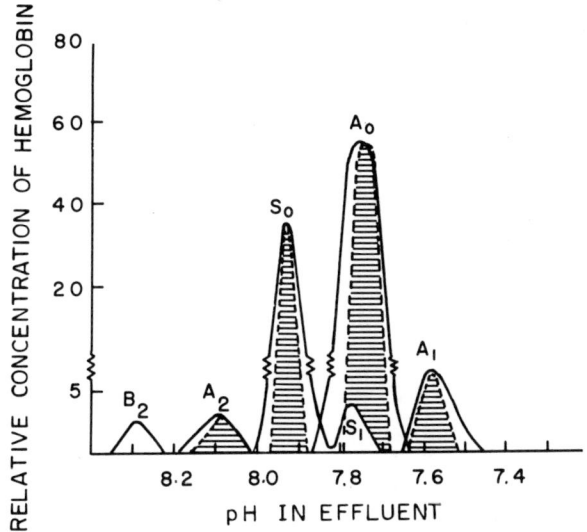

Figure 6.3. Comparison of the pH values of elution of six human hemoglobin components observed in a large number of chromatograms. For details, see text. (From Dozy, Kleihauer, and Huisman, 1968, with permission.)

A comparison between the effluent pH values of several α-, β-, or δ-chain abnormal human hemoglobins and their relative mobilities in starch-gel electrophoresis is given in Figure 6.4. The direct relationship between the pH values of elution and the electrophoretic mobilities of these components was established. With only a few exceptions, such as Hb A_2 and Hb GS (or $\alpha_2^G \beta_2^S$), Hb A_2 and Hb C, Hb A_o and Hb S_1, Hb F_o and Hb A_1, a complete separation of the various hemoglobin types in the various hemolysates can be obtained. Such a comparison may serve as a guide for the selection of the proper pH gradient to be applied to DEAE-Sephadex chromatography.

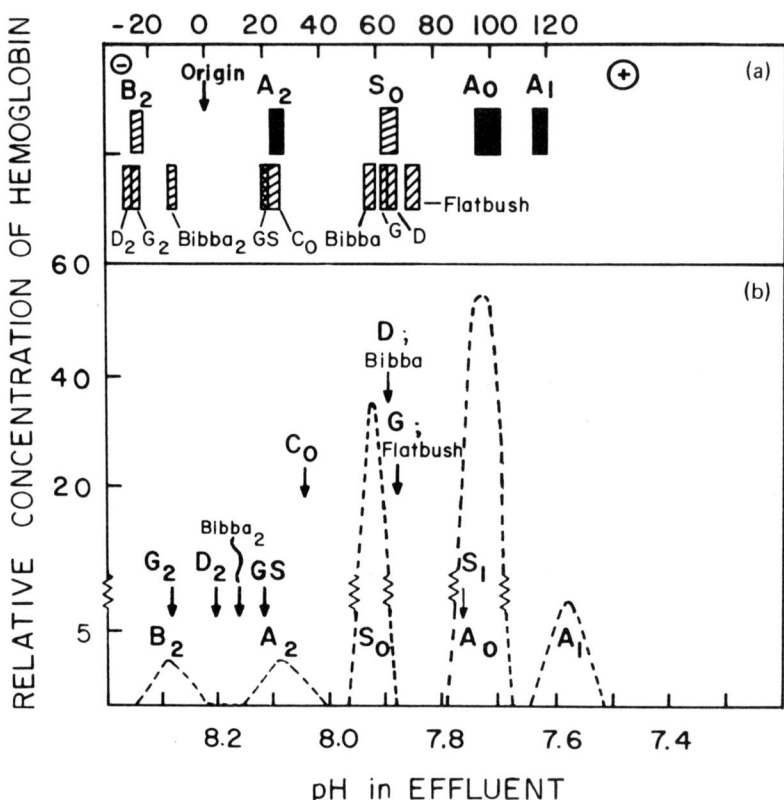

Figure 6.4. Comparison of the pH values of elution of several
human adult hemoglobin types by DEAE-Sephadex chromatography (b),
and of their relative mobilities in starch-gel electrophoresis (a).
(From Dozy, Kleihauer, and Huisman, 1968, with permission.)

III.   CHROMATOGRAPHY ON DEAE-CELLULOSE

The description of this procedure is essentially identical to that
given by Abraham *et al.* (1976-1977) with some modification.

A.   Procedure

1.   Developers

These are 0.2 M in glycine (15 g liter^{-1}) and contain 100 mg KCN
liter^{-1}. The pH of this glycine-KCN solution (Developer A) varies

between 7.7 and 7.9 and is not adjusted. All other developers are prepared from this solution by dissolving NaCl in different quantities; again, the pH of these developers is not adjusted. The compositions of the most commonly used developers are given in Table 6.1; the developers are best identified by their millimolar NaCl concentrations.

2. Preparation of Ion Exchanger

The anion exchanger DEAE-cellulose known as DE-52, microgranular and preswollen, is manufactured by Whatman, Ltd. (Great Britain) and is marketed in the United States by Whatman, Ltd. (Clifton, NJ).

About 100 g of this material is suspended in 300 ml of Developer A and stirred for 10 min with a magnetic stirrer. After settling, the supernatant is decanted and replaced by fresh developer. This procedure is repeated twice whereafter fresh developer is added so that the volume of the supernatant is about the same as that of the settled anion exchanger. The pH of the suspension is adjusted to 7.8 with 2 M HCl under continuous and rather vigorous stirring. The ion-exchanger suspension is stored in a stoppered bottle at room temperature.

3. Pouring and Equilibration of
   the Column

The DE-52 suspension is poured in sections until a 1 X 25 cm column is obtained. Equilibration is not necessary.

Table 6.1 Glycine-KCN-NaCl Developers Used in DE-52 Chromatography

Developer	NaCl		KCN	Glycine
	(mol liter^{-1})	(g liter^{-1})	(mg liter^{-1})	(mol liter^{-1})
A	0	0	100	0.2[a]
A-5	0.005	0.293	100	0.2
A-10	0.010	0.585	100	0.2
A-20	0.020	1.170	100	0.2
A-30	0.030	1.755	100	0.2
A-40	0.040	2.340	100	0.2
A-60	0.060	3.510	100	0.2
A-200	0.200	11.700	100	0.2

[a]0.2 M glycine = 15 g.

4. *Development of the Chromatogram*

Hemolysate is prepared from a freshly collected blood sample, and an appropriate volume which contains 20-30 mg hemoglobin is diluted with 1 ml distilled water and dialyzed for 24 hr at 4°C against Developer A which has been diluted with an equal volume of distilled water.

After this sample has been applied on top of the column and allowed to drain in, the top 5-mm portion of the ion exchanger is gently stirred. The hemoglobin which is attached to the inner wall of the glass tube is carefully washed down with 0.5 ml Developer A. Developer A is also layered over the ion exchanger, and the glass tube is filled with this solution. The column is connected to a 500-ml constant-volume mixer which is filled with Developer A-5. A continuously increasing NaCl gradient is obtained by introducing Developer A-30 into the mixing flask. A flow rate of 15-18 ml hr^{-1} is maintained with a pump while fractions are collected at 20-min intervals. About 24 hr later as determined by experience, Developer A-30 is replaced by Developer A-60. When a slower gradient is needed, Developer A-20 may be used in lieu of the Developer A-30; 24 hr later this developer is replaced by Developer A-40 which, in turn at a later time, is replaced by Developer A-60. If an electrophoretically fast-moving variant such as Hb N or Hb I is present, the gradient is changed after 48 hr by replacing the Developer A-60 by the Developer A-200. Chromatograms are usually completed in 48 hr or in about 140 fractions (720-860 ml), except when an extended gradient is used. The chromatogram is developed at room temperature; the recovery of hemoglobin approaches 100%.

5. *Reequilibration and Reuse*

Although reequilibration of the DE-52 column has not been very successful, it is possible to regenerate the resin that has been accumulated from several columns. After the material has been thoroughly washed with Developer A-200 to remove all residual hemoglobin, it is washed extensively with distilled water until chloride ions are absent in the effluent as tested with silver nitrate. Finally, the

material is equilibrated 3-5 times with Developer A. Regenerated
DE-52 has slightly altered chromatographic properties and is not
very suitable for analytical chromatography. However, it may be
used with success for the isolation of hemoglobins with isoelectric
points that are rather different from that of Hb A, such as Hb S,
Hb C, Hb E, Hb N-Baltimore, Hb I, etc.

B. Examples of Chromatography on
    DEAE-cellulose

Figures 6.5 to 6.7 give numerous chromatograms which illustrate the
great utility of this relatively simple procedure. The chromatograms of Figure 6.5 concern some electrophoretically slow-moving
variants. Chromatogram (a) shows the separation of the hemoglobins
from a person with the β-chain variant Hb C ($\alpha_2\beta_2$ 6 Glu $\rightarrow$ Lys) and
the δ-chain variant Hb A_2' ($\alpha_2\delta_2$ 16 Gly $\rightarrow$ Arg). Six hemoglobin
zones can be identified; Hb A_2 does not separate from Hb C, but the
minor hemoglobins C_1 and A_1 can be readily quantitated. Chromatogram (b) refers to the hemoglobin of a heterozygote for Hb Saskatoon
($\alpha_2\beta_2$ 63 His $\rightarrow$ Tyr) which elutes as a broad zone between the hemoglobins A_2 and A_o. Chromatogram (c) illustrates the complete separation of the various hemoglobins from a person with sickle cell
trait. The last chromatogram of Figure 6.5 (d) shows the complete
separation of all hemoglobins (including Hb F) in the blood from
an Hb Lepore-Washington heterozygote; the relatively low value of
Hb A_2 is characteristic of this condition.

The three chromatograms of Figure 6.6 concern electrophoretically fast-moving variants, and show a (nearly) complete separation
of the variant and the normal hemoglobins. Although the substitu-
tion in Hb Malmö ($\alpha_2\beta_2$ 97 His $\rightarrow$ Gln) (6.6a) hardly affects the
electrophoretic properties, the chromatographic separation of Hb
Malmö and Hb A and of the minor hemoglobins (A_1 and Malmö$_1$) on this
system is nearly complete, but Hb F (if present) would cochromato-
graph with Hb Malmö. The α-chain variant Hb J-Paris (α_2 12 Ala $\rightarrow$
Asp β_2) (6.6b) does not seem to separate from the normally occurring minor Hb A_1 component, but the minor variant J_2 (α_2 12 Ala $\rightarrow$

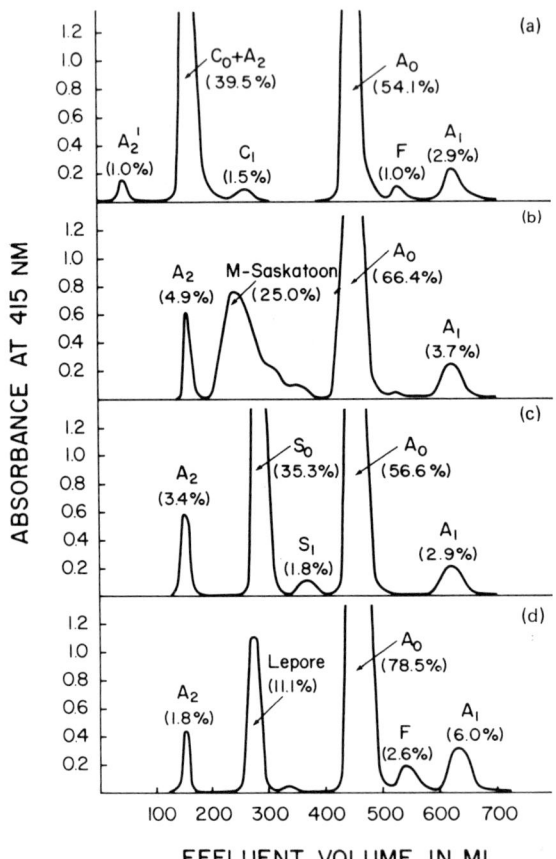

Figure 6.5. Chromatographic separation of some electrophoretically slow-moving variants on columns of DEAE-cellulose. For details, see text.

Asp δ_2) clearly separates from Hb A. Hb N–Baltimore ($\alpha_2\beta_2$ 95 Lys → Glu) (6.6c) separates completely from the hemoglobins A and A_1, and its quantitation in a red cell hemolysate can readily be made by this procedure.

Figure 6.7 illustrates the separations of complex mixtures of hemoglobin components that are present in persons who are heterozygous for an α-chain variant and heterozygous (or homozygous) for the β-chain variant Hb S ($\alpha_2\beta_2$ 6 Glu → Val). The eight hemoglobins

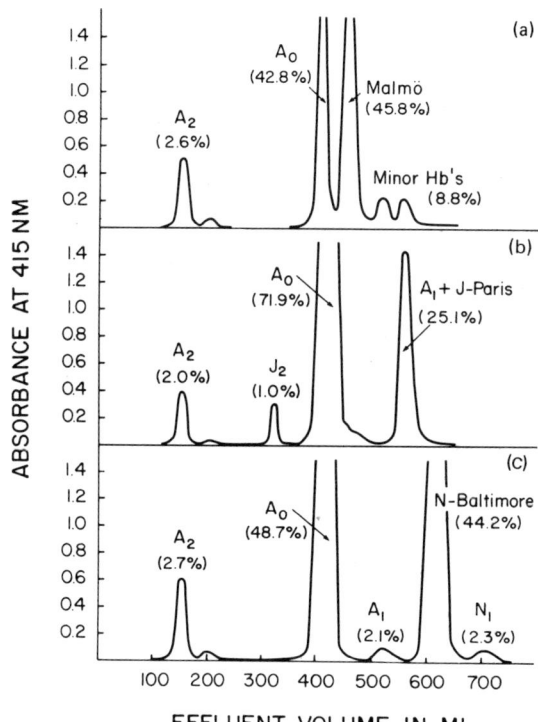

Figure 6.6. Chromatographic separation of some electrophoretically fast-moving variants on columns of DEAE-cellulose. For details, see text.

$(G_2 = \alpha_2{}^G\delta_2;\ A_2 = \alpha_2\delta_2;\ SG = \alpha_2{}^G\beta_2{}^S;\ S = \alpha_2\beta_2{}^S;\ G = \alpha_2{}^G\beta_2;\ A = \alpha_2\beta_2;$ $F = \alpha_2\gamma_2;$ and $A_1 = A$ with blocked β-chain amino termini) that are present in a patient with the Hb G-Georgia-sickle cell trait combination (G-Georgia = α_2 95 Pro → Leu β_2) were completely separated, and quantitation of each of these zones was readily possible (6.7a). A comparable chromatogram was obtained from a patient with the Hb G-Philadelphia-sickle cell trait combination (G-Philadelphia = α_2 68 Asn → Lys β_2) except that Hb A_2 and Hb SG (= $\alpha_2{}^G\beta_2{}^S$) were eluted as one zone (6.7b). The third chromatogram (6.7c) shows the separation of the hemoglobins from a patient with sickle cell anemia who also was heterozygous for Hb G-Philadelphia, and the fourth chromatogram

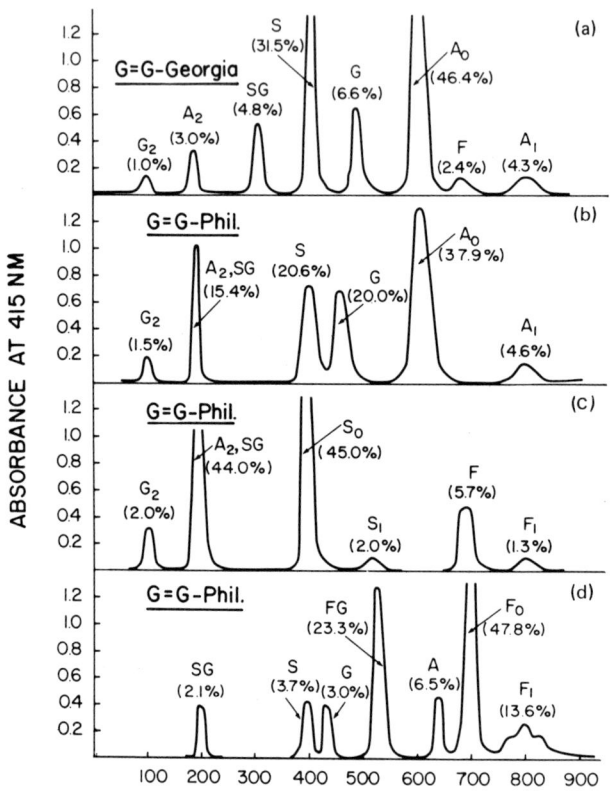

Figure 6.7. Chromatographic separation of the hemoglobins of persons with combinations of an α-chain variant and a β-chain variant on columns of DEAE-cellulose. For details, see text.

(6.7d) illustrates similar results but from a newborn who has the Hb G-Philadelphia-sickle cell trait combination. The separation of all hemoglobins is complete except for that of Hb A_2 and Hb SG $(= \alpha_2^G \beta_2^S)$; Hb FG $(= \alpha_2^G \gamma_2)$ elutes between Hb G $(= \alpha_2^G \beta_2)$ and Hb A and, thus, quantitation of the total amount of γ-chain-containing hemoglobins is possible in the newborn.

C. Comments

In the authors' laboratories, this newly developed procedure has largely replaced the DEAE-Sephadex procedure because it is simpler and faster. Moreover, because the hemoglobins move down the column as compact zones, the separation of hemoglobins in complex mixtures is better. An additional advantage is the improved separation of the hemoglobins A, F, and A_1; thus, the different elution rates of the hemoglobins A_1 and F make it possible to quantitate Hb F in samples which contain over 2% of Hb F. A detailed discussion of this procedure for the quantitative determination of Hb F is in Chapter 9.

The results in Table 6.2 demonstrate how highly reproducible are the data from this procedure. The information from the eight chromatograms was obtained within only 2 days after the collection of the blood.

D. Other Procedures of DEAE-cellulose
   Chromatography

1. *Chromatography with Phosphate-NaCl*
   Developers

This procedure, which was developed in the early 1960s (Huisman and Dozy, 1962a), used an amorphous form of DEAE-cellulose (Selectacel-DEAE, type 40) and dilute sodium phosphate developers (0.005-0.009 M) which contained KCN (0.01%) and variable amounts (0.0-0.3 M) of NaCl.

Table 6.2 Accuracy of the Quantitation of Hb A_2 and Hb S in a Blood Sample of an Hb S Heterozygote by DEAE-cellulose Chromatography

	Hb A_2	Hb S
Number of analyses	8	8
Mean (%)	3.51	40.12
Range (%)	3.31-3.66	39.59-40.96
Standard of the mean	0.04	0.49
Coefficient of variation	0.04	0.012

Two types of elution systems were developed; one in which the pH of the developers was kept constant but the NaCl concentration gradually increased, and a second in which the pH value also gradually decreased from 8.6 to 6.0. An example of the first system and of the chromatographic separation that can be obtained is given in Figure 6.8. This chromatogram was developed with the use of a 10-chamber gradient system whose chambers were filled with Na-phosphate-NaCl developers as indicated. Although this method is no longer in use, it has served as a very useful basis for the development of other procedures. The main disadvantages were the marginal separation of several hemoglobins and the asymmetric zones which promoted overlapping between the zones.

The method was also the basis for the first attempt to quantitate Hb A_2 chromatographically in several samples at the same time (Huisman, 1961; Huisman and Dozy, 1961). After DEAE-cellulose has been equilibrated with 0.005 M sodium phosphate buffer pH 8.6 with 100 mg KCN liter^{-1}, a 1 X 15 cm column is packed and 10-25 mg hemoglobin is chromatographed. Elution of the Hb A_2 fraction is made with an 0.01 M sodium phosphate buffer pH 8.6 (with 100 mg KCN liter^{-1}) at a flow rate of 40-50 ml hr^{-1}. After the elution of the Hb A_2 fraction (about 40 ml effluent is required), the column is mounted above a volumetric flask and the remaining hemoglobin is eluted with 0.01 M sodium phosphate buffer pH 6.0 to which 0.3 M NaCl is added. The percentage of Hb A_2 may then be calculated. Modified chromatographic procedures for the quantitation of Hb A_2 are described in detail in Chapter 7, Section V.

2. *Chromatography with Tris-HCl*
    Developers

This modification of the DEAE-cellulose chromatographic procedure (Efremov and Huisman, 1974) uses 0.05 M Tris-HCl developers (with 100 mg KCN liter^{-1}) with different pH values. Because the separation of hemoglobins A_o, F_o, A_1, and F_1 is greatly improved, this type of DEAE-cellulose chromatography is unusually useful for the isolation of Hb F from large volumes of blood.

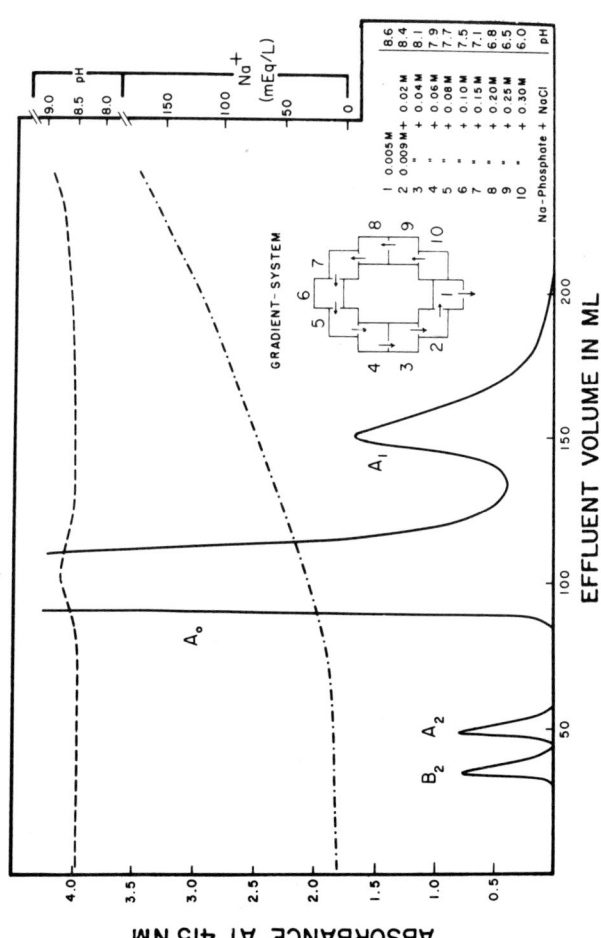

Figure 6.8. Chromatographic separation of the hemoglobins of an Hb B₂ heterozygote on a column of DEAE-cellulose; ---, pH values of the effluent; —·—·—, Na⁺ concentrations of the effluent. (From Huisman and Dozy, 1962a, with permission.)

About 100 g DE-52 is suspended in 500 ml Tris-HCl buffer, pH 8.5, and the suspension is stirred slowly for 1 hr. After settling, the supernatant is removed and discarded, and the ion exchanger is resuspended in the same amount of this developer. When this procedure has been repeated once more, the equilibrated ion exchanger is

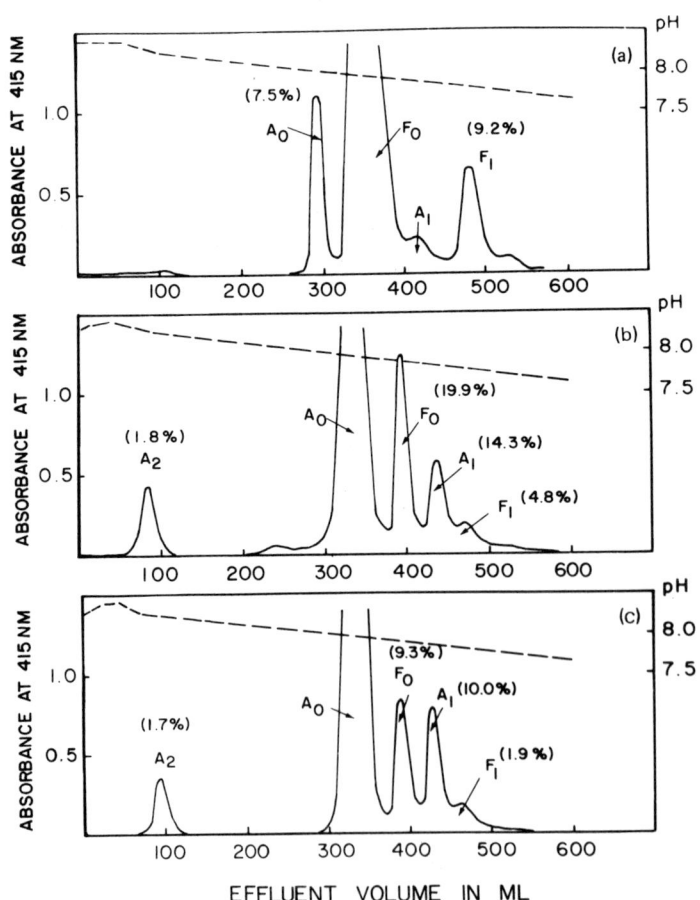

Figure 6.9. Chromatographic separation on DEAE-cellulose of the hemoglobins from a cord-blood sample (a), from a subject with an HPFH heterozygosity (b), and from a patient with Fanconi's anemia (c) with Tris-HCl developers. Broken lines represent pH values of effluents. (From Efremov and Huisman, 1974, with permission.)

stored at room temperature as a suspension of 2 parts settled material and 1 part supernatant solution.

A 0.9 X 55 cm column is packed in sections and equilibrated overnight at room temperature with Developer pH 8.3 at the rate of 30 ml hr^{-1}. The sample is 40-50 mg hemoglobin that has been dialyzed overnight at 4°C against Developer pH 8.3. The column is connected with a 250-ml constant-volume mixer which contains Developer pH 8.3. A continuously decreasing pH gradient is obtained by supplying Developer pH 7.9 to the mixer. The flow rate of the column is maintained at 18-24 ml hr^{-1} with a pump. The chromatogram is developed at room temperature. After 400-450 ml developer has passed through the column (that is, after about 24 hr), Developer pH 7.9 in the funnel is replaced by Developer pH 7.5, and the elution is continued for another 24 hr.

Figure 6.9 depicts the chromatographic separation of the hemoglobin components in three freshly prepared hemolysates which contained different amounts of Hb F. The new features of the chromatograms are the position of Hb F which is eluted between the major Hb A_o and the minor Hb A_1 components, and that of the minor Hb F_1 which is eluted last. However, considerable overlap is observed, and isolated hemoglobins F_o and F_1 from chromatograms of samples containing low percentages of these fetal hemoglobins often contain some Hb A_o and Hb A_1, whereas a small amount of Hb F_o may still be present in the Hb A_1 zones. The chromatographic quantitation of Hb F is discussed in detail in Chapter 9.

IV. COMMENTS ABOUT DEAE-SEPHADEX AND DEAE-CELLULOSE CHROMATOGRAPHY

The main purpose of this section is to compare these chromatographic procedures with each other and with electrophoretic procedures.

Application of a pH gradient, provided this gradient is produced with buffers of reasonably high molarity (for example, 0.05 M in the Tris-HCl gradient system for DEAE-Sephadex and DEAE-cellulose chromatography), results in the separation of many hemoglobins which often appear in the chromatogram as single peaks of rather symmetrical shape. Therefore, the use of Tris-HCl developers of reasonably high molarities is a great improvement over that of phosphate developers of low molarity (0.005-0.009 M) to which variable amounts of NaCl are added.

The selection of the glycine-KCN-NaCl developing system in chromatography on DEAE-cellulose combines the need of a solution with a relatively high molarity with that of a NaCl gradient which has been most useful in both anion and cation exchange chromatographic systems. This new chromatographic procedure is characterized by the appearance of symmetrical hemoglobin zones which are considerably more compact than the rather diffuse zones in DEAE-Sephadex chromatography. The NaCl gradient functions according to the common principle that the chloride ions are constantly competing with the anionic groups of the protein (COO^-) for association with the cationic groups of the anion exchanger $[(C_2H_5)_2-\overset{+}{N}H-C_2H_4-OR]$. On the other hand, the sodium ions may bind to the anionic groups of the proteins and thus compete with the cationic groups of the anion exchanger. This dual function may make the NaCl gradient more efficient than a decreasing pH gradient (thus, an increasing HCl gradient). Although developers are used in this procedure without adjusting the pH, the pH of the anion exchanger suspension needs to be adjusted rather precisely between 7.8 and 7.9. When the pH of the suspension is adjusted to anywhere between 7.5 and 7.7, the separation of electrophoretically slow-moving hemoglobins is greatly impaired and these hemoglobins are in part eluted within the void volume of the column.

The rates of elution of many hemoglobins are directly related to the electrophoretic mobilities of the proteins (Huisman and Dozy, 1965). A direct relation between the pH values of elution, when plotted on a logarithmic scale, and the relative mobilities of the various hemoglobin types in starch-gel electrophoresis has been found. It is of interest that two distinct relations have been observed: the first is a linear relation between the chromatographic and electrophoretic properties of the human hemoglobin types of the

$\alpha_2\beta_2$ series (Hb A and variants) and the $\alpha_2\delta_2$ series (Hb A_2 and variants), and the second is the same relationship of human Hb F and its γ- and α-chain abnormal variants. Because human fetal hemoglobins are always eluted at a lower pH value than is expected from electrophoretic observations, it suggests that additional properties of structurally different hemoglobins contribute to the adsorption of these proteins by the anion exchanger and, therefore, to their elution rates.

Similar relationships have also been observed for DEAE-cellulose chromatography with glycine-KCN-NaCl developers. Again, the elution rates of most hemoglobins are directly related to their electrophoretic mobilities. Figure 6.10 shows the relationship between the

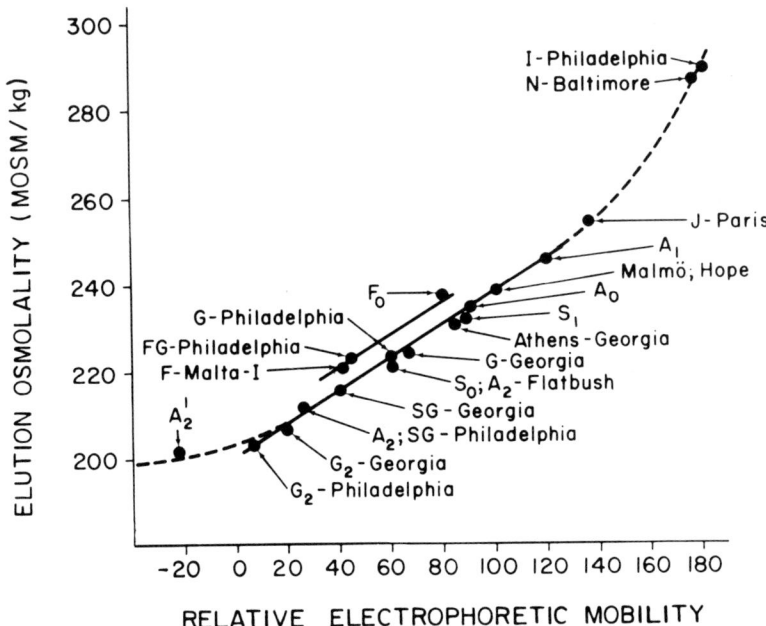

Figure 6.10. The relationship between the elution osmolalities by DEAE-cellulose chromatography in the glycine-KCN-NaCl system and the relative mobilities in starch-gel electrophoresis at pH 9.0. The mobility of Hb A was taken as 100. (From Abraham *et al.*, 1976-1977, with permission.)

osmolalities at which hemoglobin zones emerge and their relative
mobilities by starch-gel electrophoresis at pH 9.0 (Abraham *et al.*,
1976-1977). A linear relationship exists for most variants with α
and β or α and δ chains; a separate linear relation is again observed
for Hb F and some of its variants. Hb A_2' (or $\alpha_2\delta_2$ 16 Gly $\rightarrow$ Arg) is
eluted in the void volume of the column probably because the osmo-
lality of the initial developer exceeds the osmolality required for
its elution. Hemoglobins I and N are eluted under conditions of
nonequilibrium because of the rapidly increasing NaCl gradient which
is used in developing the chromatograms.

I. INTRODUCTION

In preceding chapters, the chromatographic procedures for the separation of hemoglobins have generally used columns that are 1 cm in diameter, 20-60 cm in length, with approximately a 40-mg sample of hemoglobin. As the various examples in the preceding and two following chapters depict, such procedures provide a detailed qualitative and quantitative view of both major and minor hemoglobin components in a sample. Considerable time and effort are required to obtain such information, and the methods are unsuited for the examination of many samples each day. Chromatography of hemoglobin on a reduced scale has been described by Huisman and Dozy (1961) and by Bernini (1969); the latter used 0.8 X 9 cm columns of diethyl-aminoethyl-(DEAE)-cellulose with Tris-phosphate buffers as a rapid means of quantitating Hb A_2. Likewise, Horton and Chernoff (1970) devised a "minicolumn" system with 0.2 X 15 cm columns of carboxy-methyl-(CM)-cellulose or DEAE-cellulose and made several applications. However, because one of the authors wished to examine qualitatively as many as 50 umbilical-cord blood samples per day for the presence of Hb S, columns were miniaturized, samples reduced in size, and time and effort lessened. In this particular instance, the reasoning toward a reduction in scale went somewhat as follows. A qualitative identification only of a low percentage of Hb S and/or Hb A in the presence of much Hb F was needed. For example, on Amberlite IRC-50 under appropriate conditions Hb S and Hb A are strongly fixed but separate, and Hb F washes through the column readily. Why not, therefore, dispense with most of the column? Why not use only a short column, which is really the top of a long column, to

carry out the Hb S-Hb A separation but permit the Hb F to wash
through quickly? Why not, therefore, also decrease column diameter
and sample size as well as duration by increased flow rate? On the
basis of such reasoning, a successful procedure with CM-Sephadex as
ion exchanger was devised (Schroeder, Jakway, and Powars, 1973).
Since that time the authors and their collaborators have devised
other such "microchromatographic" methods. These efforts have produced a variety of qualitative and quantitative microchromatographic
procedures that encompass the normal and many abnormal human hemoglobins. Use is made of both CM-cellulose and DEAE-cellulose. The
time required usually is about 2 hr and many simultaneous determinations may be done. Qualitative procedures can easily be used quantitatively and conditions can be modified for special separations.

II.   COMMON TECHNICAL ASPECTS OF
      MICROCHROMATOGRAPHY

These comments are made to eliminate repetition in the following
sections of this chapter and to describe aspects and useful apparatus that are common to all procedures.

In most procedures, the column is formed in a Pasteur pipet
that has an internal diameter of about 0.5 cm. The length can be
as great as 6-7 cm in a Pasteur pipet. If a column perhaps as long
as 20 cm is desired in a special experiment, glass tubing of appropriate size is drawn out and constricted at one end. The ion exchanger is held in these tubes by a cotton plug in the constricted
end. It is important that only a wisp of cotton be used and that
it be tamped *lightly, not tightly* into the constriction. If much
cotton or tight tamping is used, the flow rate may be almost nil
even with appreciable liquid heads. When the column is poured in
a plastic drinking straw (see Section VI), the cotton should be
tamped into a 2-3 mm layer on the special fitting.

When short columns of 5-6 cm are used in a Pasteur pipet, it
is convenient to adjust the ratio of settled resin to supernatant
fluid so that a column of appropriate length is formed from one

complete filling of the pipet with slurry. As the fluid drains out
and the ion exchanger settles, no concern need be felt that the column will "go dry" and be ruined. In fact, it should go dry before
the sample is added. After the column has gone dry, the sample
should then be applied within a few minutes. Columns may be prepared and stored for future use. For storage, some fluid is left
above the column. The bottom is closed with a sealer of the type
that is used for microhematocrit tubes or with a length of plugged
plastic tubing and the top with a cork.

It is inconvenient to hold these small chromatographic columns
in ordinary clamps. An inexpensive holder for multiple columns may
be made by placing rubber bands over thumbtacks that are spaced
about 5 cm apart in two parallel rows about 3 cm apart on a piece
of wood. When the Pasteur pipet is placed behind the two parallel
rubber bands, it is conveniently held for column pouring and sample
application. The board itself may be held to a laboratory support
with a common three-finger clamp.

Development may be done in several ways. After the sample has
been applied and the tube above the column has been filled with developer, the tube may be attached by rubber tubing to a funnel and
placed in a funnel rack. Appropriate developer is then placed in
the funnel. The length of the funnel stem will determine the liquid
head and influence the flow rate through the column. Another convenient apparatus is a manifold with 5 or 10 outlets that are supplied by a single reservoir whose position determines the liquid
head. The outlets have rubber tubing with pinch clamps and individual chromatograms may easily be attached and detached. In commercial kits for the determination of Hb A_2 (see Section V), the
column is prepared in a plastic tube with an expanded section above
the column itself to form a reservoir for a specified volume of
developer. All microchromatographic procedures are done at room
temperature.

Chromatograms that are of qualitative interest only may be
preserved for considerable periods, especially in the cold, simply
by closing both ends to prevent drying out.

The pH to which the ion exchanger is equilibrated is critical
in these procedures. However, different designs of electrodes and
pH meters react differently to a stirred suspension of charged particles. Some are sensitive to the rate of stirring. Consequently,
the same suspension may not have the same pH reading on different
instruments. Although a pH is specified in the descriptions below,
a user may have to adjust to a slightly different pH in order to
obtain comparable results. This topic is discussed in some detail
in Huisman *et al.* (1975) and Section V.

In most of the microchromatographic procedures, blood itself
(instead of hemolysate from which cell debris has been removed) may
be used as the sample after appropriate additions of water or solutions to produce hemolysis. Because a volume of blood rather than
an amount of hemoglobin is specified, the actual amount of hemoglobin that goes into the determination will vary with the packed-cell
volume; therefore, two samples of the same percentage composition
may appear to differ significantly on the chromatograms.

III. QUALITATIVE DETECTION OF
      HEMOGLOBINOPATHIES AT BIRTH

An initial procedure on 0.5 X 6 cm columns of CM-Sephadex in Pasteur
pipets (Schroeder, Jakway, and Powars, 1973) which was used successfully for more than 10,000 determinations (Powars, Schroeder, and
White, 1975) suffered from the disadvantage that the zones were
diffuse on the translucent CM-Sephadex column. The method that is
described below produces very distinct zones on the white background
of a CM-cellulose column with only 20% of the amount that the CM-
Sephadex procedure used. The description below is taken from
Schroeder *et al.* (1975).*

---

*Much of this section is paraphrase or direct quotation from
Schroeder *et al.* (1975) with permission.

A. Procedure

1. Developers

Developer 1 is 0.03 M Tris-HCl/0.025 M NaCl/0.01% KCN at pH 6.1 and
is prepared with 3.63 g Tris, 1.46 g NaCl, 0.1 g KCN, plus 6 M HCl
to pH 6.1 in 1 liter.

Developer 2 is 0.03 M Tris-HCl/0.04 M NaCl/0.01% KCN at pH 6.3
(3.63 g Tris, 2.34 g NaCl, 0.1 g KCN, plus 6 M HCl to pH 6.3 in 1
liter).

2. Preparation of Ion Exchanger

The microgranular, preswollen CM-cellulose designated CM-52 and
manufactured by Whatman (Clifton, NJ) is used. It is prepared in
Developer 1 above as described in Chapter 5, Section V.A.2. The
final slurry should have equal volumes of settled ion exchanger and
supernatant solution.

3. Pouring of the Column

A 6-cm column is poured in a Pasteur pipet as described in Section
II. No equilibration is necessary. Each column requires about 1 g
ion exchanger.

4. Development of the Chromatogram

The sample is prepared by mixing 0.02 ml blood, 0.3 ml Developer 1,
and 0.2 ml 0.004 M maleic acid, allowing at least 5 min at room
temperature before application to the column.

Most of the liquid above a poured column is drawn off and any
remaining liquid is allowed to flow into the column. The sample is
then carefully applied without disturbing the top of the column and
allowed to flow in. Finally, the tube above the column is filled
with Developer 1 and the column is attached with rubber tubing to
a manifold or funnel. The chromatogram is developed with 40-50 ml
Developer 1 with a liquid head of about 50-60 cm to give a flow
rate of approximately 20-25 ml hr^{-1}. If a strongly fixed hemoglobin
is present after this development, the solvent above the column is

replaced by Developer 2 and 6-8 ml are passed through. The column cannot be used again.

B.   Examples of Detection of Abnormal
     Hemoglobins in Cord Bloods

The hemoglobin zones are compact and easily visible against the white background of the CM-cellulose column.

The typical appearance of the chromatogram after development with Developer 1 is shown by the left chromatogram of each pair in Figure 7.1 for cord bloods from AA, AS, SS, and AC children.   The Hb F in the sample remains partly on the column, but some is in the effluent.   Hb A forms a zone with limits between about 10 and 25 mm from the top, whereas Hb S is in a 2-3 mm zone that has barely moved from the top and Hb C remains in the upper 0.5-1.0 mm.

Hb S and Hb C may really be distinguished by the fact that Hb C is very strongly fixed, whereas Hb S moves slightly down the column. The conclusion is substantiated by change to Developer 2 with the result that is seen in the right chromatogram of each pair in Figure 7.1.   Thus the few milliliters of Developer 2 elute Hb F, move Hb A to the middle of the column, remove Hb S from the top, but leave Hb C virtually unmoved.   The SC and CC conditions may readily be distinguished from sickle cell anemia by such use of Developer 2.

Hb E, although it behaves much like Hb C in starch-gel electrophoresis, can be distinguished from Hb C by this procedure.   It is less strongly fixed than Hb C and moves somewhat like Hb S with which it could be confused.   Addition of more than the prescribed volume of Developer 2 in comparison with a control sample permits the distinction to be made.

C.   Comments on Technical Aspects

Spurious effects will be observed if the pH of the sample is not carefully controlled.   The chosen proportion of blood and reagents has been carefully determined in order to provide a final pH of 6.1 or less.   If, for example, more blood is used in relation to the

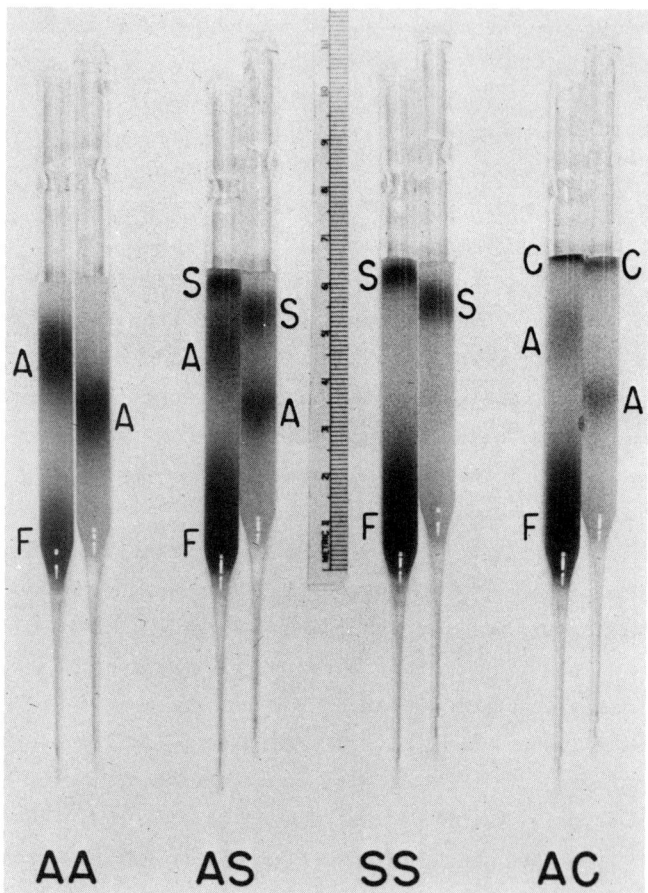

Figure 7.1. Chromatograms of the hemoglobins in cord bloods of
infants with the indicated conditions. After completion of chromatography with Developer 1, the chromatograms have the appearance
in the left-hand member of each pair. Hb S and Hb C are distinguished by their differential movement with Developer 2 as shown
in the right-hand member. (From Schroeder *et al.*, 1975, with permission.)

reagents, the initial pH will be too high and a portion of the Hb F
will pass rapidly through the column. Although the final appearance
of the chromatograms will be about normal, the rapid passage of part
of the Hb F will incorrectly suggest a heterogeneity of Hb F.

      Because Tris has only meager buffering action at pH 6.1, fluc-
tuations in the pH of ion exchanger and solutions may occur and can
be detected by abnormal chromatographic behavior. Truly buffered

conditions may be obtained by substituting bis-Tris which has a pK_a of about 6.5. Virtually identical behavior results if Developer 1 is constituted with 0.03 M bis-Tris-HCl/0.03 M NaCl/0.01% KCN at pH 6.2. The greater cost of bis-Tris may not warrant its routine use.

Although samples are normally refrigerated, they may be maintained at room temperature for long periods without adversely affecting the results. Thus, whole blood samples that were maintained at room temperature for as long as 1 year gave results that were comparable to the initial run on fresh material. AA samples produced no extraneous material that might be confused with Hb S or Hb C, nor did Hb A or Hb S disappear from an AS sample.

The method is so sensitive that satisfactory results are still obtained with a much-reduced sample size. Thus, the normal sample of 20 μl blood for a 0.5 X 6 cm column can be reduced to 4 μl. If the column is only 0.3 X 6 cm, 2 μl blood yields easily detectable zones of Hb A and Hb S in cord blood. Because Hb A and Hb S in cord blood are each of the order of 5-10%, it means that Hb A and Hb S are detectable in 0.1-0.2 μl blood of adults with sickle cell trait.

IV. QUALITATIVE DETECTION OF
     HEMOGLOBINOPATHIES IN ADULTS*

Although the quantitative microchromatographic procedures that are discussed in later sections of this chapter were designed for the rapid determination of specific hemoglobins, it became obvious through their use that they also could be of wide application in a qualitative way. This section describes how microcolumns either of CM-cellulose or DEAE-cellulose may be used in the qualitative study of some infrequently observed hemoglobin variants. Not only may electrophoretically fast-moving hemoglobins such as J ($\alpha_2\beta_2$ 16 Gly → Asp) and N ($\alpha_2\beta_2$ 95 Lys → Glu) be detected but Hb S ($\alpha_2\beta_2$ 6 Glu → Val)

---

*Much of this section is paraphrase or direct quotation from Abraham *et al.*, 1977a, with permission.

can be distinguished from Hb D ($\alpha_2\beta_2$ 121 Glu → Gln) and Hb C ($\alpha_2\beta_2$ 6 Glu → Lys) from Hb E ($\alpha_2\beta_2$ 26 Glu → Lys) and Hb O-Arab ($\alpha_2\beta_2$ 121 Glu → Lys).

A. CM-cellulose with bis-Tris-NaCl Developers

1. *Procedure*

a. Developers

Developer C-1 has 6.28 g bis-Tris, 2.34 g NaCl, and 0.1 g KCN liter^{-1} and is adjusted to pH 6.2 with concentrated HCl. It is 0.03 M bis-Tris/0.04 M NaCl/0.1% KCN.

Developer C-2 has 6.28 g bis-Tris, 2.92 g NaCl, and 0.1 g KCN liter^{-1} and also is adjusted to pH 6.2 with concentrated HCl. It is 0.03 M bis-Tris/0.05 M NaCl/0.01% KCN.

b. Preparation of the Ion Exchanger

Microgranular preswollen CM-cellulose (CM-52; Whatman, Clifton, NJ) is equilibrated with Developer C-1 (or C-2 if appropriate).

c. Development of the Chromatogram

A 0.5 X 15 cm column of CM-52 that has been equilibrated with Developer C-1 (or C-2) is poured either in a glass tube or a plastic drinking straw (see Section VI). The sample is prepared from 0.02 ml blood, 0.2 ml 0.004 M maleic acid, and 0.3 ml 0.05% saponin (Calbiochem) in Developer C-1 (or C-2). At least 5 min should be allowed for hemolysis. After the sample is applied, the chromatogram is developed with Developer C-1 (or C-2) at a flow rate of about 10 ml hr^{-1}. The two developers differ only in molarity of sodium chloride and are chosen in specific cases to provide a reasonable movement of the hemoglobins.

2. *Examples*

This method has proved to be useful for the separation of hemoglobin types with closely identical electrophoretic mobilities. For instance, when the blood of an individual with SD (probably D-Los Angeles, $\alpha_2\beta_2$ 121 Glu → Gln) disease is chromatographed with 60 ml Developer C-1, two zones are apparent on the column although not completely separated (Figure 7.2d). Similarly, when the blood of a patient with Hb S ($\alpha_2\beta_2$ 6 Val → Glu) and Hb Deer Lodge ($\alpha_2\beta_2$ 2 His → Arg) is chromatographed with 25 ml Developer C-2, the hemoglobins occupy positions shown in Figure 7.2e. On starch-gel electrophoresis at pH 9, Hb S and Hb Deer Lodge produce a band that is broader in the anodal direction but distinct from Hb S alone. This slight

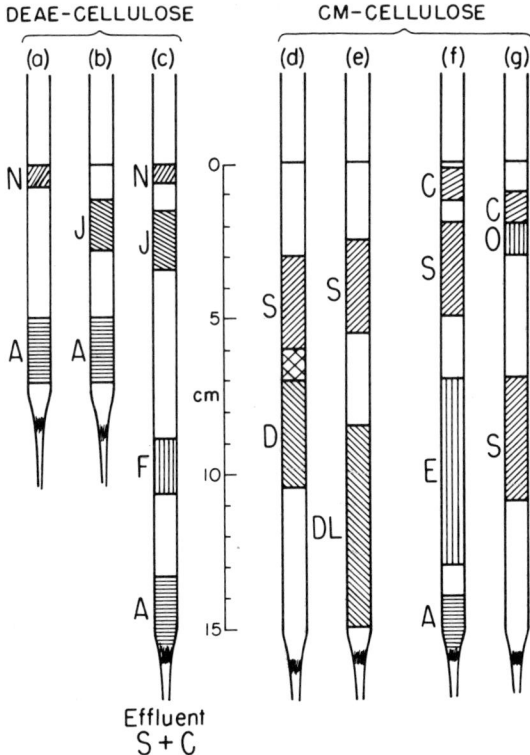

Figure 7.2. The positions and separations of some common and uncommon hemoglobins on DEAE-cellulose and CM-cellulose under various conditions that are described in the text. The limits of the zones are depicted as sharp whereas, in fact, they are diffuse. (From Abraham *et al.*, 1977a, with permission.)

difference in charge is no doubt responsible for the better chromatographic separation of Hb S and Hb Deer Lodge as compared to that of
Hb S and Hb D.

The separation of Hb C ($\alpha_2\beta_2$ 6 Glu → Lys) and Hb E ($\alpha_2\beta_2$ 26
Glu → Lys) is also readily accomplished. When AC, AS, and AE samples are chromatographed on parallel columns and developed with 50
ml Developer C-1, the final positions that are shown in composite
in Figure 7.2f are obtained.

3.   Comments

The distinction between the SS and SD genotypes or between the CC
and CE or CO genotypes is commonly done by citrate-agar electrophoresis in which hemoglobins A, S, and C take distinctive positions, but hemoglobins D and E behave like Hb A, and Hb O moves
between the hemoglobins A and S. Microchromatography under these
conditions provides more definite distinctions because Hb D and Hb
Deer Lodge not only separate from Hb S and, likewise, Hb E and Hb
O-Arab ($\alpha_2\beta_2$ 121 Glu → Lys) from Hb C but they do not mimic Hb A.

The movement of hemoglobins in these developers is very dependent upon NaCl concentration, and the examples provided illustrate
the altered movements that change in NaCl concentration brings about.
No doubt by increase in NaCl concentration, Hb C and Hb O could be
more completely separated (Figure 7.2g). Because of the sensitivity
to sodium chloride concentration, parallel chromatograms of known
and unknown hemoglobins provide a more accurate comparison of chromatographic behavior.

B.   DEAE-cellulose with Glycine-NaCl
     Developers

1.   Procedure

    a.   Developers

Three developers are used. Developer A (see Table 6.1) is 0.2
M glycine/0.01% KCN (15 g glycine and 0.1 g KCN liter^{-1}) where Developers A-15 and A-20 are the same with, in addition, 0.015 M NaCl

(0.88 g liter^{-1}) and 0.02 M NaCl (1.17 g liter^{-1}), respectively. The pH is about 7.6 and is not adjusted.

b. Preparation of the Ion Exchanger

Microgranular preswollen DEAE-cellulose (DE-52; Whatman, Clifton, NJ) is suspended in Developer A and the pH is adjusted so that the separation of Hb A$_2$ and Hb S is optimal as described in Section V.B.

c. Development of the Chromatogram

An 0.5 X 8 cm column is poured from equilibrated DE-52 in a Pasteur pipet or an 0.5 X 15 cm column in a glass tube or plastic drinking straw (see Section VI). One drop of blood or undialyzed hemolysate is mixed with 8 drops of water. If blood is used, hemolysis should be allowed to occur for at least 5 min before the sample is applied to the column. After the sample has been applied, Developer A is used at a flow rate of about 15 ml hr^{-1} for the shorter column (10 ml hr^{-1} for the longer column) until the effluent equals 5 ml. Developer A is then replaced by Developer A-15 or A-20 and about 15 ml of one of these for the shorter column and about 60 ml for the longer column is then allowed to flow through.

2. Examples

Chromatography on the shorter column is the procedure for the quantitative determination of Hb A$_2$ (see Section V.B). Thus, Hb A$_2$ as well as any Hb C will pass through virtually unretarded. By the time that 5 ml Developer A has been used, Hb S will have moved to the middle of the column, while any Hb A, Hb F, and electrophoretically fast-moving hemoglobins at alkaline pH will adhere to the top of the column. After development with 15 ml Developer A-20 is complete, Hb A and Hb F are at the bottom, Hb J is in the middle, and Hb N is still at the top of the column. Figures 7.2a and 7.2b show the results when the method is applied to AN-Baltimore ($\alpha_2\beta_2$ 95 Lys → Glu) and AJ-Baltimore ($\alpha_2\beta_2$ 16 Gly → Asp) samples.

Experiments with other fast-moving hemoglobins give somewhat different results. Hb K-Woolwich ($\alpha_2\beta_2$ 132 Lys → Gln), for instance, moves more rapidly down the microcolumn than Hb N, but more slowly than Hb J. Hemoglobin Hope ($\alpha_2\beta_2$ 136 Gly → Asp) elutes between hemoglobins A and J. On the other hand, Hb N and Hb I (α_2 16 Lys → Glu β_2) are indistinguishable, and Hb A and Hb F also do not separate on these relatively short columns. When a complex artificial mixture of hemoglobins C, S, A, F, J, and N is applied to the longer column and developed as indicated above, the hemoglobins emerge as well-separated zones (Figure 7.2c).

3. Comments

The procedures will clearly distinguish many known or possible heterozygosities for hemoglobins. Thus far, the hemoglobins A, S, C, F, J, N, I, K-Woolwich, and Hope have been studied. Of the possible combinations, only that of N and I would not be identified, and Hb F and Hb Hope might be confused on microchromatographic evidence alone. It is important to note that the samples of blood or hemolysate should be no older than three weeks or alteration products of Hb A (that is, Hb A_1) will interfere with the identification of Hb F or even possibly Hb J.

4. Application to Mass Testing

In order to test the applicability of microchromatography to mass testing for hemoglobinopathies under adverse conditions, the procedure for the determination of Hb A_2 was used qualitatively in Ghana (Efremov and Huisman, 1975; Efremov *et al.*, 1974b). The facilities were limited and the working conditions were rather extreme. For instance, a pH meter and a spectrophotometer were not available, and high temperatures (33°C) and high humidity (80-90%) often hampered the procedure.

For this study, the anion exchanger suspension was prepared in Georgia and shipped to Ghana where a check by microcolumn chromatography was made with an AS blood sample as a control to determine whether Hb A_2, Hb S, and Hb A were adequately separated. If not, the

the pH was adjusted in increments until the separations were satisfactory.

The Ghana survey was made exclusively with fingertip blood
which was collected on filter paper; the papers were immediately
sealed in plastic bags and refrigerated until tested. To prepare
the hemolysate, 1-cm squares of filter paper were placed in test
tubes with 10 drops of water. The resulting hemolysate was placed
on the top of a 0.5 X 4 cm column of DE-52 which had been equilibrated with Developer A and adjusted to appropriate pH. The types
of hemoglobin were identified by their rates of elution with Developer A. As many as 80-100 samples could be analyzed daily.

During a 3-month period in which 2350 bloods were examined by
this method under these suboptimal conditions, the following major
technical problems were encountered:

1. The pH of the buffer changed due to the growth of mold.
2. The high temperature and the high humidity hampered the
    evaluation of the eluted fractions; sometimes a fraction
    such as Hb C would elute, sometimes it would not.
3. The lack of a pH meter hampered the proper equilibration
    of the resin. Although the accuracy of the identification
    was unaffected, time was spent in checking the ion exchang-
    er with a control sample from an Hb S heterozygote until
    the proper separations were obtained.

Among the 2350 subjects, 15.4% had sickle cell trait and 9.4%
had Hb C trait. Moreover, 15 patients with sickle cell anemia, 7
with homozygous Hb C disease, and 22 with the SC condition were
discovered. Thus, it appears that the most common variants were
readily identified by the microchromatographic procedure.

The samples were also sent to Georgia for examination by starch-
gel electrophoresis. In those instances in which a meaningful electrophoretic examination could be made, the electrophoretic and microchromatographic results agreed in 99% of the comparisons. The electrophoretic methods detected cases with elevated levels of Hb F or

variants of Hb A$_2$ which this microchromatographic procedure was not designed to do.

V. QUANTITATIVE DETERMINATION OF HEMOGLOBIN A$_2$

A. Hemoglobin A$_2$ with Tris-HCl Developers

This procedure is the same as that described by Efremov and collaborators (1974a).

1. Developers

These are prepared from 1.0 M Tris-HCl buffer, pH 8.8. The four developers are 0.05 M Tris, contain 100 mg KCN liter^{-1}, and are titrated with 4 M HCl to the desired pH values of 8.35, 8.30, 8.20, and 7.00, respectively. The pH values of the 8.35, 8.30, and 8.20 developers should be checked regularly and, if they alter during storage as has sometimes been observed in buffer that was stored in open containers, they should be adjusted with the 0.05 M Tris-HCl buffer, pH 8.5, which contains 100 mg KCN liter^{-1}.

2. Preparation of the Ion Exchanger

About 100 g of the anion exchanger DE-52 is equilibrated by mixing with 500 ml 0.05 M Tris-HCl buffer, pH 8.5, which contains 100 mg KCN liter^{-1}. After nearly all material has settled, the supernatant solution and fine particles (fines) are removed, and the equilibration procedure is repeated at least three times. The equilibrated DE-52 is stored as a slurry in a covered container; the supernatant volume of buffer is about 0.7 of that of the settled anion exchanger.

3. Pouring of the Column

When columns in Pasteur pipets are filled with slurry of the proportions given above, one filling will produce a final column of about 6-7 cm in length. If the prepared column is not used immediately, it may be stored after filling the remaining space above the column with 0.05 M Tris-HCl buffer, pH 8.5, and capping the tube.

Larger columns (0.5 X 15 cm) are used for the analyses of samples
with Hb S; these columns are prepared in the same way as the shorter
ones.

4. Development of the Chromatograms

Blood can be collected by venipuncture in EDTA vacutainer tubes
(Becton-Dickinson, Rutherford, N.J.) and by fingerstick either in
heparinized microhematocrit capillary tubes (Sherwood Medical Indus-
tries, Inc., St. Louis, Mo.) or on filter paper (Whatman 3 MM).
Usually two capillary tubes or a 1-in square filter paper with 1 or
2 drops of blood should be available. Prior to chromatography, 1
drop of hemolysate (1 drop of an 8-12 g dl^{-1} hemolysate from a 0.2
ml laboratory pipet or disposable Pasteur pipet) is diluted with 6
drops of distilled water and 1 drop of a KCN (2 g dl^{-1}) solution
(referred to as hemolysate sample). It is also possible to use the
content of about 1/2 of one completely filled heparinized capillary
tube which is mixed with 8 drops of distilled water and 1 drop of
KCN (2 g dl^{-1}) solution (whole blood sample). The filter paper
(one falling drop of blood collected on Whatman No. 2 or No. 3 MM
filter paper contains 3-6 mg hemoglobin) is cut into small pieces
and submerged for at least 15 min in about 0.5 ml of an elution
reagent that contains 0.1% tetrasodium ethylenediaminetetraacetate
and 1 mM KCN (filter paper sample). Each of these three hemoglobin
solutions is chromatographed without dialysis or additional treatment.

  If the packed column has been stored, it is carefully uncapped
and all but a 5-mm layer of buffer is removed. When necessary, the
top of the column is leveled by tapping the side carefully. The
sample is applied to the column and after the solution has settled
into the column, pH 8.30 developer is pipetted gently on top. Finally, the tube is attached to the funnel or manifold. The effluent
is collected in a test tube which is calibrated at 10 ml or in a
volumetric flask. Hb A$_2$ is visible on the column as a band about
2-5 mm wide and is eluted in about 1 hr during which time 8-9 ml

effluent are collected. The remaining hemoglobin is eluted with
the pH 7.00 developer in about 10 ml and is collected in a calibrated test tube or volumetric flask. After the volumes of the two
effluent fractions have been adjusted to 10 and 25 ml, respectively,
the absorbance at 415 nm is measured.

The procedure for the 0.5 X 15 cm column has these modifications. The collection of the Hb A_2 zone which is eluted with pH
8.35 developer is started when this hemoglobin has reached the lower
part of the column. The adjusted volume of the eluted Hb A_2 zone
is 10 ml. The zone of Hb S is eluted with pH 8.20 developer, and
the remaining hemoglobin with pH 7.00 developer; the adjusted volume
of each of these zones is 25 ml. This modified procedure requires
about 5 hr to complete.

5. Calculation of Percentages

Calculations are made as follows (see also Chapter 2, Section VII):
For a 0.5 X 5 cm column:

$$\% \text{ Hb } A_2 = \frac{A_{A_2} \times 100}{A_{A_2} + 2.5 A_R}$$

For a 0.5 X 15 cm column:

$$\% \text{ Hb } A_2 = \frac{A_{A_2} \times 100}{A_{A_2} + 2.5 A_S + 2.5 A_R}$$

$$\% \text{ Hb S} = \frac{2.5 A_S}{A_{A_2} + 2.5 A_S + 2.5 A_R}$$

where A = absorbance at 4.5 nm and R = remaining hemoglobin.

6. Examples of Results Obtained
*   with this Technique*

The procedure has been used in several large surveys. Often the
data have been compared with those obtained by other chromatographic

Table 7.1 The Percentages of Hb A_2 Determined by Different Procedures

Method	Normal adult		β-Thalassemia trait	
	n^a	Average	n^a	Average
Cellulose acetate	48	2.4 $(1.0-4.2)^b$	13	5.9 $(3.9-8.1)^b$
DEAE-Sephadex	40	2.65 $(1.8-3.2)^b$	15	5.6 $(3.9-6.7)^b$
DEAE-cellulose (macro)	124	2.3 $(1.5-3.0)^b$	45	5.0 $(3.5-6.3)^b$
DEAE-cellulose (micro)	5362	2.5 $(1.3-3.7)^b$	326	4.8 $(3.7-7.0)^b$

[a] Number of samples

[b] Range of values

techniques, such as macrochromatography on columns of DEAE-Sephadex or DEAE-cellulose, and by a cellulose-acetate electrophoretic method in which Hb A_2 and Hb A were eluted from the unstained strips. Some of these results are listed in Table 7.1. The data are comparable and differentiation of normal adults and β-thalassemia heterozygotes by their Hb A_2 values can readily be made with either one of these four procedures.

Probably the largest microchromatographic survey with the intent to detect β-thalassemia heterozygotes was conducted in Yugoslavia (Efremov and Huisman, 1975). Of the 6000 persons who participated in this study, many were school children from six cities in Macedonia, and others were parents or other relatives of known β-thalassemia homozygotes. Figure 7.3 depicts the Hb A_2 levels for the two groups; nearly no overlap was observed. Figure 7.4 presents most of the same data but in relation to the total hemoglobin level of each subject. These and other data demonstrate that the β-thalassemia heterozygote can readily be identified by the elevated level of Hb A_2 alone although other measurements such as Hb level and mean

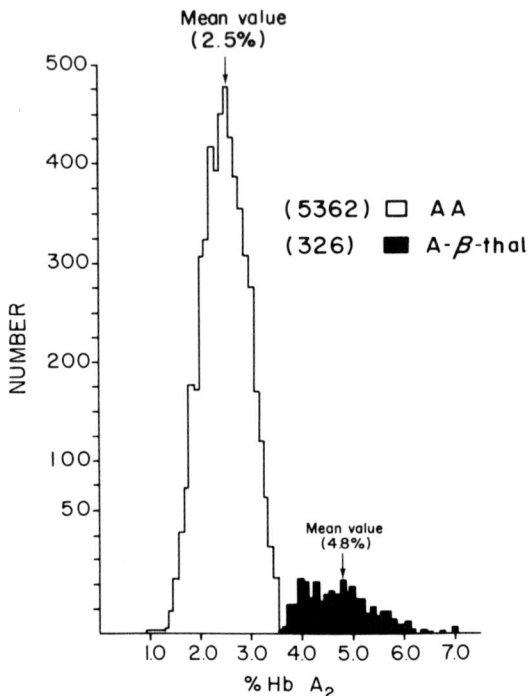

Figure 7.3. The levels of Hb A_2 in normal adults and in β-thalassemia heterozygotes (Yugoslavian survey). (From Efremov *et al.*,
1974a, with permission.)

corpuscular volume (MCV) are helpful parameters. When the hemoglobin from all these persons was also studied by starch-gel electrophoresis, nearly 30% of the β-thalassemia heterozygotes were mis-
diagnosed.

7. *Comments on Technical Aspects*

    a. The Effect of pH Changes

    Trace amounts of Hb A are eluted with Hb A_2 if the pH is 8.25
and lower; thus, the pH of the first developer must be between 8.30
and 8.35. When the Hb A_2 zone moves very slowly or appears to be
diffuse, the pH of the developer may be too high or the molarity of
the buffer may be in error.

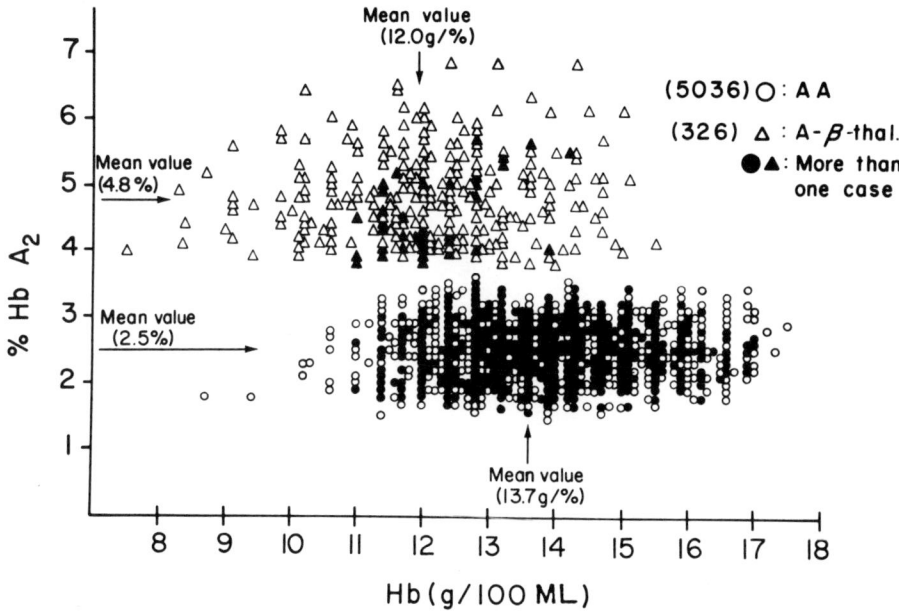

Figure 7.4. The relationship between the total hemoglobin levels and the levels of Hb A_2 (Yugoslavian survey). (From Efremov *et al.*, 1974a, with permission.)

b. The Effect of Overloading

When hemoglobin in excess of 7-8 mg is applied, the Hb A_2 component is usually slightly contaminated with Hb A. However, if samples contain 2 mg hemoglobin or less, low absorbance readings affect the precision of the Hb A_2 quantitation.

c. The Effect of Flow Rate

The flow rate can be increased to 60 ml hr^{-1} without affecting the resolution of Hb A_2 and Hb A. This high flow rate can be obtained by extending the liquid head (measured from the bottom of the column) above a freshly poured column from the commonly used 20-100 cm. The Hb A_2 zone is broadened considerably at such a high flow rate but is eluted in less than 10 min.

d. Storage of Samples

Samples of whole blood can be stored for 2-3 weeks at 4°C and
in the dark. Addition of 1-2 drops of diluted (1:1000) merthiolate
solution is recommended. Blood collected on filter paper should be
stored in a plastic envelope at 4°C, but should be analyzed within
1 week.

e. Shipment of Samples

Shipment should preferably be done under refrigeration and by
air so that the receiving laboratory will be able to analyze the
samples soon after collection.

f. Types of Samples

The three types of hemoglobin solutions (that is, conventional
hemolysates, hemolyzed whole blood, and hemoglobin extracted from a
filter paper sample) give comparable results provided the analyses
are made within 7 days after collection. The filter paper sample
appears to be the least satisfactory, and because it has the widest
range of normal values, could, on occasion, lead to the diagnosis
of a normal person as a β-thalassemia heterozygote.

g. Normal Values

An average value of 2.5% was found for over 6000 normal adults
with a range of 1.3-3.7%. The average value for more than 300 previously diagnosed β-thalassemia heterozygotes was 4.8% with a range
of 3.7-7.0%.

Data on samples containing Hb S gave the following average
values and ranges:

AS:	2.8%; range 1.6-3.9%	(n = 303)
SS:	3.05%; range 1.7-4.4%	(n = 194)
S-β-thalassemia	4.9%; range 3.8-6.1%	(n = 35)
S-HPFH:	2.1%; range 1.1-2.6%	(n = 20)

These values were obtained with the large 0.5 X 15 cm columns and
show that Hb A_2 determinations are useful for a differential diagnosis of the SS, S-β-thalassemia, and S-HPFH conditions.

B. Hemoglobin A$_2$ with Glycine-KCN
   Developers

This modified procedure has been described by Huisman *et al.* (1975)
and allows the quantitation of Hb A$_2$ without interference from Hb S
(or Hb D) in the sample. Elution of Hb A$_2$ is much less sensitive
to minor change in pH of the developer, but is greatly dependent on
the pH of the ion exchanger.

1. Developers

Developer A is 0.2 M glycine/0.01% KCN (15 g glycine and 0.1 g KCN
liter^{-1}); the pH of this solution is not adjusted. Developer A-200
is 0.2 M glycine/0.2 M NaCl/0.01% KCN (15 g glycine, 11.7 g NaCl,
and 0.1 g KCN liter^{-1}). These developers are the same as or related
to those in Table 6.1. Although Developer A-200 is used as the
second developer, the nature of the second developer is not important, but it needs to be concentrated enough to remove Hb A and Hb
F quickly; for instance, it may be replaced by physiological saline.

2. Preparation of the Ion Exchanger

About 100 g DE-52 is mixed with sufficient Developer A to give a
total volume of 400 ml. After 10 min of mechanical stirring, the
ion exchanger is allowed to settle and the supernatant solution is
discarded and replaced with fresh Developer A. This procedure is
repeated twice, whereafter the ion exchanger is suspended in fresh
Developer A. After stirring for 10 min, the pH of the stirred suspension is adjusted to 7.6 with about 8-10 meq HCl. Next, a test
chromatogram is made with a fresh sample from a Hb S heterozygote.
If Hb A$_2$ does not emerge in the first 3-4 ml and Hb S in the next
15-20 ml Developer A, the pH of the suspension is readjusted under
stirring to pH 7.5 and another test is made. If necessary, further
reduction of pH and testing is made until the "correct" pH is reached.
Because various electrodes react differently to a stirred suspension
of charged particles (see Section II), adjustment of pH in the manner
described may be difficult. In an alternate procedure, the suspension is allowed to settle and the pH is measured in the quiet super-

natant solution. A test chromatogram is made as above. If adjustment is necessary, acid is added to the stirred suspension but pH
again is measured in the supernatant solution. Testing is made
until the correct pH is attained. In one of the author's laboratory,
this pH is 7.1-7.2. Subsequent preparations of DE-52 should be adjusted to the pH as determined by either method. There are slight
differences in the properties of different lots of the DE-52 ion
exchanger, and the amount of HCl to bring the suspension to a given
pH will vary slightly. Some supernatant fluid is removed so that
the volumes of settled ion exchanger and supernatant fluid are about
equal. Suspensions of DE-52 have been used for as long as 3-4 weeks
without alteration of properties. However, occasionally an increase
in pH on aging has been observed; thus, if the behavior of the hemoglobin on the ion-exchange column changes, the pH of the suspension
should be tested and readjusted, if necessary. On the other hand,
if the pH has decreased or has been adjusted to too low a value, it
cannot satisfactorily be raised to the proper value.

3. Pouring of the Column

A 5-6 cm column is prepared in a Pasteur pipet as described above.

4. Development of the Chromatogram
   in the Absence of Hb S

One drop of blood or the contents of one microhematocrit tube or 1
drop of hemolysate is mixed with 8 drops of water; the hemolysis is
carried out for 5 min. After this hemolysate has been applied to
the top of the column and allowed to run in, Developer A is added
above the column. The column is next attached to a funnel filled
with Developer A. Hb A_2 (or Hb C) will be eluted in 3-4 ml. Developer A is replaced with Developer A-200 to elute the Hb A. The
effluent with Hb A_2 is diluted to 5 ml and that with Hb A to 25 ml.
The percentage of Hb A_2 is calculated from the absorbances at 415
nm of the two effluents.

5. *Development of the Chromatogram*
   in the Presence of Hb S

If the percentage of Hb A_2 only is desired, chromatography in the presence of Hb S is made exactly as in its absence; thus, Hb A_2 is eluted in the first 3-4 ml with Developer A, and next Hb S and any remaining Hb are eluted by change to Developer A-200. If the percentages of Hb A_2, Hb S, and the remaining Hb are desired, the procedure is modified as follows: Hb A_2 is eluted in the first 3-4 ml with Developer A, Hb S then emerges in the next 15-20 ml of the same Developer A, whereafter the remaining Hb is removed with Developer A-200. The effluent with Hb A_2 is diluted to 5 ml and that with Hb S and the remaining hemoglobin each to 25 ml. The percentages of the fractions are calculated from the absorbance at 415 nm of the three effluents (see Chapter 2).

6. *Examples of Results Obtained*
   with this Technique

Another large survey in Georgia made use of this technique. The samples were collected from normal adults, known β-thalassemia heterozygotes from various racial and ethnic backgrounds, from Hb S heterozygotes, and from patients with sickle cell disease or with sickle cell-β-thalassemia. The total number of persons was 2827. The data, given in Figure 7.5, demonstrate that a β-thalassemia heterozygote and a normal adult can readily be distinguished. Of some interest is the observation that β-thalassemia heterozygotes of Dutch origin have considerably higher Hb A_2 values. The results also show that for the majority of cases the percentage of Hb A_2 can be used to differentiate between the SS and the S-β-thalassemia conditions.

Figure 7.6 illustrates the relationship between the MCV values and the Hb A_2 levels. In no instance was there overlap between the normals and the β-thalassemia heterozygotes.

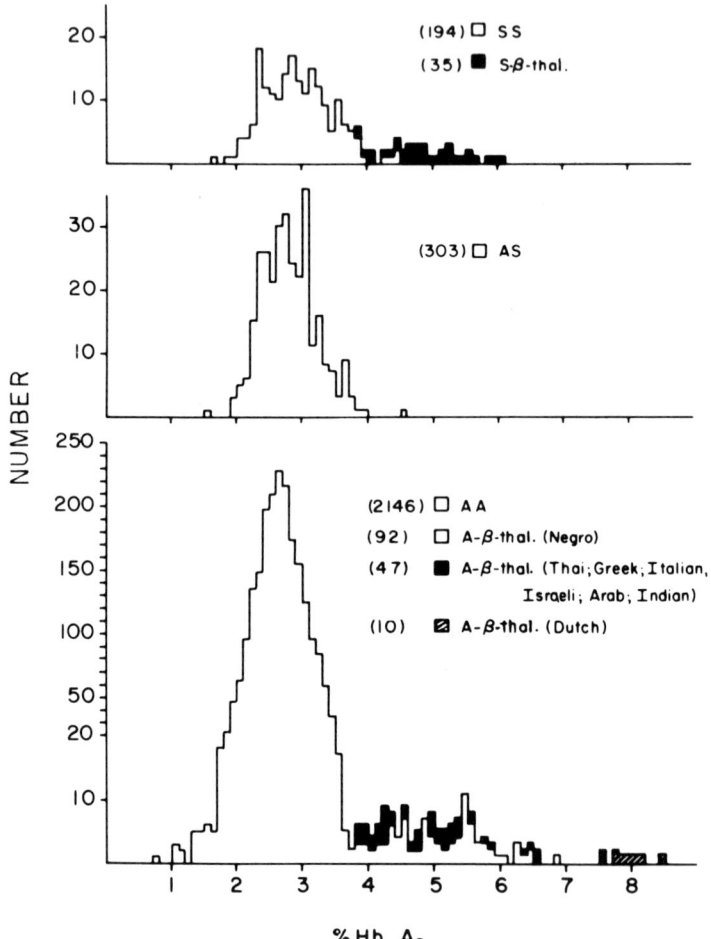

Figure 7.5. The levels of Hb A₂ in normal adults, in β-thalassemia heterozygotes, in Hb S traits, and in patients with sickle cell anemia and Hb S-β-thalassemia (Georgia survey). (From Efremov *et al.*, 1974a, with permission.)

The quantities of Hb S in the 303 Hb S heterozygotes mentioned in Figure 7.5 varied between 27 and 39%; similar values have been found by the macro DEAE-cellulose chromatographic procedure (Huisman, 1977).

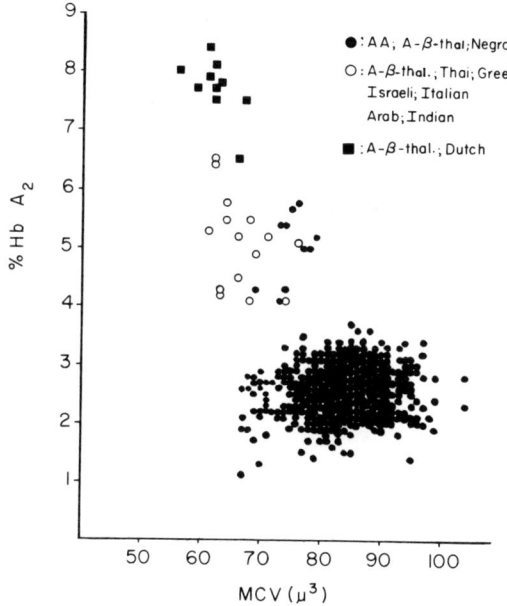

Figure 7.6. The relationship between the MCV and the levels of Hb A$_2$ (Georgia survey). (From Efremov *et al.*, 1974a, with permission.)

7. *Comments on Technical Aspects*

Statements in Section V.A about the effects of overloading the column, as well as storage, shipment, and types of samples to be analyzed are also applicable to this modified procedure. The main advantages of this new technique are the decreased sensitivity of minor changes in the pH of the developer and the fact that the same type of column can be used for samples with or without Hb S.

Figure 7.7 compares the data that were obtained with this procedure (the glycine method) with those that were obtained with the method of Section V.A (the Tris method). The correlation is excellent and differentiation between normal adults and β-thalassemia heterozygotes can be made by both procedures without difficulty. The Hb A$_2$ values in the Hb S heterozygotes are also about the same with the two techniques.

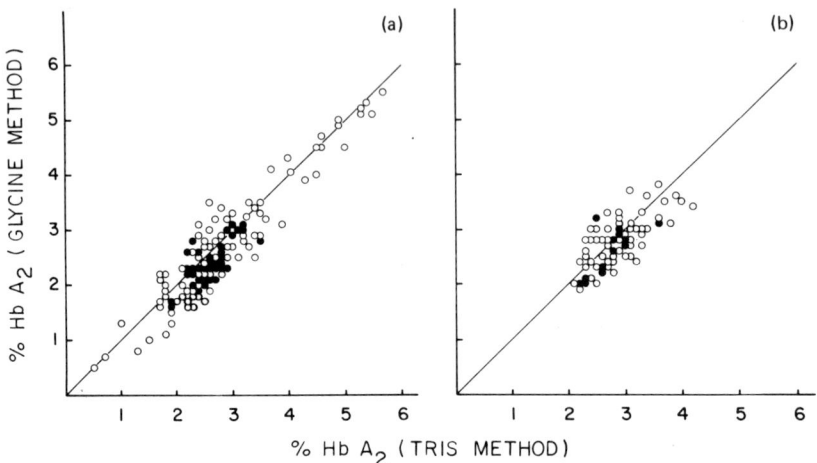

Figure 7.7. A comparison of the percentages of Hb A_2 by two different microchromatographic procedures. (a) Analyses of samples from
153 normal persons and 15 persons with a β-thalassemia trait. (b)
Analyses from 86 samples from Hb S heterozygotes. Open circles:
single cases; closed circles: more than one case. (From Huisman
et al., 1975, with permission.)

A similar comparison of levels of Hb S in Hb S heterozygotes
also gave comparable results. However, more reproducible data are
obtained if these samples are analyzed soon after collection, and
preferably within 3 days.

C. Hemoglobin A_2 with Modified
Glycine-KCN Developers

This method was recently described by Schleider, Mayson, and Huisman
(1977) and has the advantage that the elution of Hb A_2 is much less
sensitive to minor changes in pH of the ion exchanger.

1. Developers

Developer A is 0.2 M glycine/0.01% KCN (15 g glycine and 0.1 g KCN
liter^{-1}); the pH of this solution is not adjusted. Developer A-18
contains 0.018 M NaCl (1.053 g liter^{-1}) and Developer A-200 contains 0.200 M NaCl (11.700 g liter^{-1}). These developers are identical with or related to those in Table 6.1.

2. Preparation of the Ion Exchanger

The ion exchanger which is DE-52 is prepared exactly as in Section V.B above except that the pH is adjusted to 7.8. The equilibrated ion exchanger is stored in a stoppered bottle at room temperature as a suspension of 2 parts settled material and 1 part supernatant solution.

3. Pouring of the Column

A 5-6 cm column in a Pasteur pipet is prepared as described above.

4. Development of the Chromatogram

One drop of conventionally prepared centrifuged hemolysate is mixed with 4 drops of Developer A and 4 drops of distilled water. Also satisfactory is the red-cell content of one centrifuged microhematocrit tube. The latter is mixed with 4 drops of distilled water, and after 5 min with 4 drops of Developer A. After either preparation has been applied to the column and allowed to run in, 0.5-1.0 ml Developer A is added to assure that all hemoglobin has been washed into the top layer of the cation exchanger. Hb A_2 is eluted with 8 ml Developer A-18 and the remaining hemoglobin with 4 ml Developer A-200. The effluents are diluted to appropriate volumes (Hb A_2 to 8 ml and Hb A to 24 ml), and the percentage of Hb A_2 is calculated from the absorbance at 415 nm of the two effluents.

5. Examples of Results Obtained with this Technique

Table 7.2 compares the results of determinations which used this procedure and the one that is described in Section V.B above. The data compare favorably, and differentiation between normal adults and β-thalassemia heterozygotes can readily be made. The new modification has the advantage of being much less sensitive to changes in pH of developer and ion exchanger, but cannot readily be applied to samples that also contain Hb S.

6. Comments on Technical Aspects

Prefilled microcolumns with appropriate reagent solutions that use this procedure are commercially available (Isolab, Inc., Drawer

Table 7.2 Comparison of Hb A_2 Values in Hemolysates of Blood from
Normal Adult and β-Thalassemia Heterozygotes as Determined by Two
Microchromatographic Methods

Condition	Procedure	Number of samples	Mean	Range
Normal	Method B[a]	99	2.6	1.8-3.2
	Method C[b]	99	2.5	1.6-2.9
β-Thalassemia trait	Method B[a]	24	4.6	4.1-5.8
	Method C[b]	24	4.45	4.1-5.4

[a]Method B is described in Section V.B and uses DE-52, pH 7.3-7.5,
with a 0.2 M glycine/0.01% KCN developer to elute the Hb A_2.
[b]Method C is described in this section and uses DE-52, pH 7.8 ± 0.05,
with a 0.2 M glycine/0.01% KCN/0.018 M NaCl developer to elute the
Hb A_2.

4350, Akron, Ohio 44321). Repeated analyses of several batches of
these prefilled columns in one of the author's laboratory have given
satisfactory results, and have shown the usefulness of the modified
technique for Hb A_2 testing on a large scale.

The samples to be analyzed should not be older than 1 week.
Moreover, statements made in Section V.A of this chapter regarding
loading of the column (either in excess of 7 mg or less than 2 mg),
storage, shipment, and types of samples to be analyzed are equally
applicable to this modified method.

VI. COMPLETE QUANTITATION BY THE DRINKING STRAW METHOD*

Microchromatographic procedures have been designed with a specific
goal such as the qualitative detection of Hbs S and C in cord blood
(Section III), the quantitative determination of Hb A_2 (Section V),

*Much of this section is paraphrase or direct quotation from
Schroeder, Pace, and Huisman, 1978, with permission.

or the qualitative detection of various hemoglobins in adults (Section IV). However, microchromatographic procedures may be made equally applicable to the simultaneous quantitative determination of many of the hemoglobin components in a blood. Such a procedure, however, requires longer columns than can be prepared in Pasteur pipets. As a result, their successive elution not only would require much solvent and time but also considerable attention on the part of the operator unless automatic fraction collecting was done. In the procedure to be described, conditions are arranged so that some components emerge early and can be collected in large fractions with minimal attention by the operator during a working day. The remainder of the chromatogram is automatically completed overnight, and some components remain on the column. Because the column is packed in a plastic drinking straw, each section that contains a hemoglobin component may be cut out with a sharp knife and eluted separately before spectrophotometry to determine the percentages.

A. Procedure

1. *Equipment*

Clear plastic drinking straws of the type that may be purchased from restaurant suppliers are used to contain the chromatographic column: their inner diameter is about 0.6 cm and their length is 20 cm. It is convenient to have Lucite fittings for the top and bottom of the straw (Figure 7.8a). Only diameter A (0.572 cm; 0.225 in) of the dimensions in Figure 7.8a is critical and may have to be varied slightly; about 20% of a given lot of straws have the wrong diameters for a given dimension of fitting. In place of such fittings, a 2-3 cm length of constricted glass tubing which is plugged with cotton and attached to the straw with rubber tubing is a somewhat less satisfactory bottom fitting; the other end of the straw is fixed with rubber tubing to a supply of solution. Any perforated object that fits the straw tightly can probably be used as a bottom fitting.

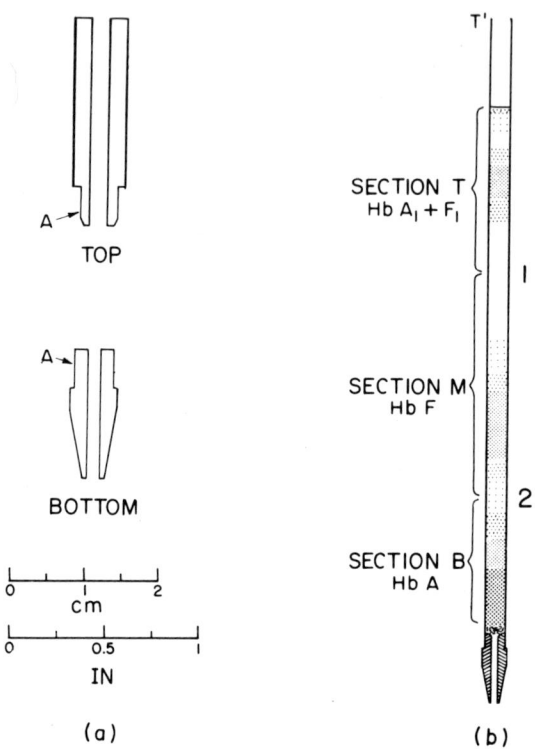

Figure 7.8. (a) Dimensions of fittings. (b) Appearance of a typical completed chromatogram. (From Schroeder, Pace, and Huisman, 1978, with permission.)

2. *Developers*

Two developers are used. Developer A is 0.2 M glycine/0.01% KCN (15 g glycine and 0.1 g KCN liter^{-1}) and Developer A-15 is 0.2 M glycine/0.015 M NaCl/0.01% KCN (15 g glycine, 0.88 g NaCl, and 0.1 g KCN liter^{-1}). The pH of these solutions is unadjusted; these developers are the same or related to those in Table 6.1.

In special cases, it may be desirable to adjust the NaCl concentration to slow down or speed up the rate of movement of a hemoglobin. For example, an electrophoretically rapid hemoglobin at alkaline pH may require increase in NaCl to 0.02, 0.03, or 0.04 M.

3. Preparation of the Ion Exchanger

DEAE-cellulose (DE-52, microgranular and preswollen; Whatman, Clifton, N.J.) is equilibrated with Developer A as described in Section V.B above and adjusted to the desired pH by either of the methods there described. The chromatograms are equally satisfactory regardless of whether the pH of the ion exchanger is not adjusted or is adjusted as low as pH 7.1 in the supernatant solution. The choice will depend on the objectives of the analysis. Usually adjustment to a pH of 7.1 is preferable. This topic is discussed in more detail below.

4. Pouring of the Column

After a bottom fitting has been attached to a plastic straw and a small plug of cotton inserted, a 17.5 cm column is poured.

5. Development of the Chromatogram

The sample which is approximately 10 mg hemoglobin (5 mg if cord blood is used) may be undialyzed hemolysate in 0.2-0.3 ml water or a hemolysate of 3 drops of blood and 12 drops of water, which is kept at room temperature for 5-10 min to allow for hemolysis. After the sample has been applied, the tube above the column is filled with Developer A-15, a top plastic fitting is inserted, and the assembly is attached to a funnel. A 60-ml portion of Developer A-15 is placed in the funnel and allowed to run through. If the initial liquid head above the bottom of the column is 20, 40, or 60 cm, the flow rate is approximately 3, 6, or 9 ml hr^{-1}. Because of the limiting amount of developer, the chromatogram may be started at any convenient time and will complete itself without attention when it runs dry.

6. Elution of the Zones

The final appearance of the chromatogram will depend somewhat on the goal of the analysis. Figure 7.8b depicts the typical final appearance of a chromatogram in which, for example, Hb F is elevated in the sample. Hb A_2 will have passed through as will part of the Hb A. These two hemoglobins may be collected separately or together. The hemoglobins that remain on the column are eluted in the following way.

After the straw has been marked at 1 and 2, which are the mid-points of the interzones (Figure 7.8b), the top fitting is removed and the straw is inserted into a 5-cm length of slightly oversize glass tubing as far as point 1. The end of the glass tubing that is positioned at point 1 should be constricted carefully so that the straw is relatively tightly held and will not flex when cut. The column is cut at point 1 and then at point 2.

After a 10-cm glass extension has been attached with rubber tubing to the top of the section B, the extension is filled with 2% KCN and the remaining Hb A is eluted and combined with the portion that has passed through the column.

The middle section M is attached at one end with rubber tubing to a 2-3 cm piece of constricted glass tubing which has a cotton plug and at the other end to an extension which is filled then with 2% KCN for elution.

Finally, a small plug of cotton and a bottom fitting are inserted into the top section T at T'. By tapping at T', the column will slide against the cotton plug. Then 2% KCN is added for elution.

An alternative method of elution may be used. After each section has been blown separately into a centrifuge tube, the inside of the straw is rinsed with 2% KCN. More KCN is added and the suspension is shaken for a few minutes. Subsequent to centrifuging, the supernatant solution is removed and the procedure is repeated. The combined supernatant solutions are used for spectrophotometry. The volume of each washing with KCN solution must be at least 10 times the volume of the centrifuged ion exchanger.

Absorbance of all fractions is read at 415 nm and percentages are calculated as usual.

B.   The General Nature of the
     Chromatogram

In a mixture that might contain Hb A_2, Hb C, Hb C_1, Hb S, Hb S_1,
Hb A, Hb A_1, Hb F, and Hb F_1 in various combinations, Hb A_2 or a
mixture of Hb A_2 and Hb C will form the first zone to move down
the column.   Hb A_2 and Hb C are the hemoglobins whose rate of movement is influenced by the pH to which the ion exchanger has been
adjusted.   If the DE-52 is equilibrated with Developer A without
pH adjustment, the pH of the supernatant solution is about 7.6.   If
the chromatogram is developed on this ion exchanger with Developer
A-15, the Hb A_2 (or Hb C) forms a narrow zone and moves slowly.   On
the other hand, if the pH is adjusted to 7.1 in the supernatant
solution, Hb A_2 (or Hb C) moves rapidly as a somewhat diffuse zone.
Thus, quantitative estimation of Hb A_2 is most conveniently done
by using ion exchanger at the lower pH.   At a flow rate of 6 ml
hr^{-1}, collection of the Hb A_2 can begin after about 1 hour and is
complete at the end of 2 hr.

     Hb S follows Hb A_2 (or Hb A_2 and Hb C) through the column.   If
the sample contains no Hb A, Hb S_1 is apparent in the position of
Hb A above and well separated from Hb S.   In the presence of Hb S
and Hb A, Hb A and Hb S_1 coincide, and analogously Hb S and Hb C_1.
Above Hb A (and/or Hb S_1) is Hb F and finally near the top of the
column will be Hb $A_1 + F_1$.

     The goal of the analysis will determine the course of the pro-
cedure.   Thus, if all components of an AS sample are to be determined, the chromatogram is started with ion exchanger at pH 7.1 at
least 2 hr before the operator will leave so that Hb A_2 may be collected separately.   Collection in the second flask is then begun
and the chromatogram is allowed to go to completion (and go dry)
without further attention.   If 55 ml rather than 60 ml Developer
A-15 is used, the Hb S will be almost completely in the second
flask.   However, some will be at the bottom of the column so that

the midpoint between Hb S and Hb A + Hb S_1 is apparent. Finally, the midpoints between Hb S and Hb A + Hb S_1, between the latter and Hb F, and between Hb F and Hb A_1 + Hb F_1 are marked, the column is sectioned, the materials are eluted, and the quantities are determined.

C. Examples

Complete quantitative determination of the components in the hemoglobin from a variety of hematological conditions was done both by macrochromatography as described in Chapter 6, Section III and by the straw method. Both procedures use DE-52 as ion exchanger and glycine-KCN-NaCl solutions as developers. Table 7.3 presents the results. The agreement of the results by the two methods is excellent and compares favorably with what might be expected from duplicate determinations by either method. The higher value for A_1 + F_1 or F_1 by the straw method is due to the complete removal of material from the top section in contrast to the tendency for trailing of the last zone and incomplete removal in macrochromatography.

In this section, the use of the straw method for the general quantitation of mixtures of hemoglobins has been emphasized. The determination of Hb F is part of such a procedure. However, Chapter 9 will discuss in detail the use of chromatographic methods, including the straw method, for the quantitation of Hb F.

D. Comments on Technical Aspects

1. pH

As mentioned above, Hb A_2 and/or Hb C are hemoglobins whose chromatographic behavior in the straw system is much influenced by the pH to which the ion exchanger is adjusted. The choice of pH for the systems will depend upon the user's objectives.

2. Concentration of NaCl

The concentration of NaCl is critical for the separation of Hb F from Hb A. If it is 0.013 or 0.017 M instead of 0.015 M, the

Table 7.3 Comparison of Quantitative Data by Macrochromatography and by the Microchromatographic Straw Method[a]

Condition	Hemoglobin composition (%)[b]
Normal adult	$A_2 = 2.4$ (2.6); $A = 92.5$ (89.0); $F = 1.2$ (2.2); $A_1 + F_1 = 3.9$ (6.2)
Normal adult	$A_2 = 2.3$ (2.5); $A = 88.7$ (84.8); $F = 2.6$ (3.1); $A_1 + F_1 = 6.3$ (9.6)
Normal adult	$A_2 = 2.5$ (2.4); $A = 90.5$ (86.1); $F = 1.2$ (2.2); $A_1 + F_1 = 5.8$ (9.4)
Normal adult	$A_2 = 2.5$ (2.7); $A = 91.7$ (89.8); $F = 1.3$ (1.5); $A_1 + F_1 = 4.5$ (6.0)
β-thalassemia trait (adult)	$A_2 = 5.2$ (5.3); $A = 87.4$ (84.8); $F = 2.4$ (2.9); $A_1 + F_1 = 5.1$ (7.1)
S trait (adult)	$A_2 = 4.2$ (3.7); $S = 35.9$ (35.0); $A = 56.0$ (55.4); $F = 1.2$ (1.6); $A_1 + F_1 = 2.7$ (3.8)
D trait (adult)	$A_2 = 2.2$ (2.2); $D = 37.4$ (37.4); $A = 47.5$ (45.1); $F = 7.6$ (8.0); $A_1 + F_1 = 5.5$ (7.3)
E trait (adult)	$A_2 + E = 32.3$ (28.9); $A = 61.5$ (62.9); $F = 0.7$ (1.8); $A_1 + F_1 = 5.3$ (6.4)
New York trait (adult)	$A_2 = 2.8$ (3.4); $A = 53.7$ (51.9); $NY = 39.5$ (39.2); $? = 4.0$ (5.5)
SS disease (adult)	$A_2 = 2.6$ (2.3); $S = 67.3$ (73.0); $F = 23.4$ (23.6); $F_1 = 6.8$ (10.6)
S-HPFH (adult)	$A_2 = 1.8$ (1.6); $S = 64.0$ (65.2); $F = 28.2$ (26.9); $F_1 = 5.9$ (6.4)
SC disease (adult)	$A_2 + C = 37.6$ (41.3); $S = 42.5$ (37.0); $F = 16.6$ (17.3); $F_1 = 3.3$ (7.0)
SC disease (cord blood)	$A_2 + C = 8.5$ (8.5); $S = 7.9$ (7.5); $F = 73.2$ (68.1); $F_1 = 10.5$ (15.9)

[a]From Schroeder, Pace, and Huisman, 1978, with permission.

[b]The first of each pair of numbers is the result from macrochromatography and those in parentheses from the straw method. In the data as given, the minor components (except for A_1 and F_1) are summed with the appropriate major component.

separation is unsatisfactory. Under slightly different conditions
in another laboratory, this parameter might have to be altered somewhat in order to achieve maximum separations.

3. *Flow Rate*

A flow rate of about 6 ml hr^{-1} is convenient. At 9 ml hr^{-1} the
zones are more diffuse and the separations worsen. Little is gained
by reducing the flow rate to 3 ml hr^{-1}.

4. *Volume of Developer*

This variable determines the final separation of the components.
Whether more or less than the recommended 60 ml is used will depend
upon the objectives of the analysis. For example, if a complete
analysis of an AS sample is desired, it is advantageous to use a
few ml less so that the interzone between Hbs A and S is still apparent in the lower column. On the other hand, if the determination
of Hb F is of main importance, a few milliliters more might be added
so that, while most of the Hb A is in the filtrate, the Hb A to Hb F
interzone is still obvious. Furthermore, in another laboratory
under slightly different conditions, the volume of developer might
have to be altered to accommodate the differences.

5. *Quantity of Sample*

For adult samples with normal or moderately reduced packed cell
volumes, the hemoglobin in three drops (about 0.07-0.08 ml) or one
microhematocrit tube is adequate. If the hematocrit is very low,
it is desirable to centrifuge and remove some plasma. Only 1 or 2
drops of cord blood should be used because the high percentage of
Hb F will prevent adequate separation from Hb A if the amount of
hemoglobin on the column is high.

6. *Storage of Samples*

If samples are refrigerated, blood may be stored for 3-4 weeks as
blood, and the percentage of Hb F will not change significantly.
On the other hand, increasing amounts of material fall in the A_1 +

F_1 region in older samples. The increase is observed within a few days in hemolysates but not for about 2 weeks with blood.

7. *Time Required*

This microchromatographic procedure requires more elapsed time than other microchromatographic methods. However, little attention is required of the operator. Routinely 12 analyses per day are easily made. The chromatograms may be started as late as 2 hr before the operator leaves for the night if Hb A_2 is to be collected separately. The chromatogram completes itself (goes dry) during the night and cutting, elution, and spectrophotometric determination then are made in the morning. If an operator devoted his time exclusively to this method of analysis, 20-25 analyses per day seem a reasonable goal.

VII. QUANTITATIVE DETERMINATION OF HEMOGLOBIN F

The microchromatographic and macrochromatographic determination of Hb F is described in Chapter 9.

I. INTRODUCTION

The preceding chapters have provided abundant evidence that the hemoglobin of the adult human, as well as that of the newborn human, is heterogeneous. This heterogeneity is, of course, compounded if hemoglobin variants are present. Hb F, the major component at birth, has the relatively minor component Hb A (15-30%) and the very minor component Hb A$_2$ ($\sim$ 0.2%), but postnatally Hb F rapidly becomes the minor component ($\sim$ 1%) of Hb A ($\sim$ 90%) and Hb A$_2$ still is minor (2-3%). The chromatographic behavior and quantitative determination of these hemoglobins have been detailed in Chapters 5, 6, 7, and 9.

However, the various chromatograms that are depicted in those chapters have minor components far in excess of these three hemoglobins; minor hemoglobins, therefore, are more than simply the changing proportions as a function of age of the individual. Although historically such minor hemoglobins were observed by Morrison and Cook (1955) and by Prins and Huisman (1956), Allen, Schroeder, and Balog (1958) were the first to detect heterogeneity in these components and to show the presence of three to five minor hemoglobins which are eluted in front of the major Hb A zone, and which have electrophoretic mobilities at pH 8.5-9.0 distinctly greater than that of Hb A$_o$. According to the order of elution, these hemoglobins have been designated A$_{Ia}$, A$_{Ib}$, A$_{Ic}$, A$_{Id}$, and A$_{Ie}$, respectively. In further studies, various properties of some of these minor hemoglobins were examined (Clegg and Schroeder, 1959; Holmquist and Schroeder, 1966; Huisman and Meyering, 1960; Jones and Schroeder, 1963b; Schnek and Schroeder, 1960). Although Schnek and Schroeder (1960) correlated these minor hemoglobins from Amberlite IRC-50 with the behavior of various bands by starch-block electrophoresis and Huisman and Meyering (1960) did a similar correlation with carboxymethyl-(CM)-cellulose chromatography, the relationship of the various minor components as they may be detected by the several kinds of cation- and anion-exchange chromatography has never been determined. Because Amberlite IRC-50 chromatography appears to be the most effective procedure for separating these minor components and

because the determination of Hb A_{Ia+b} and Hb A_{Ic} appears to be of
value in the evaluation of diabetes, this chapter will discuss modifications of Amberlite IRC-50 chromatography of Hb A_{Ia+b}, Hb A_{Ic},
and other minor hemoglobins.

Hemoglobin A_{Ic} which accounts for the major portion of the
minor hemoglobins, has a hexose moiety (1-amino, 1-deoxy-fructose)
at the amino-terminus of the β chains by virtue of formation of a
Schiff base and an Amadori rearrangement (Bookchin and Gallop, 1968;
Bunn *et al.*, 1975; Holmquist and Schroeder, 1966; Koenig, Blobstein,
and Cerami, 1977). The glycosylation of Hb A to produce Hb A_{Ic} appears to be a slow, nonenzymatic posttranscriptional event which
takes place over the life span of the red blood cell (Bunn *et al.*,
1976). Hb A_{Ic} comprises some 4-6% of the total hemoglobin in normal human red cells but is increased, as is Hb A_{Ia+b}, in patients
with overt diabetes mellitus (Trivelli, Ranney, and Lai, 1971). It
has now been suggested that control of diabetic hyperglycemia can
be monitored by measuring Hb A_{Ic} or Hb A_{Ia+b+c} (Koenig *et al.*, 1976).
Investigation of this aspect of Hb A_{Ic} has now reached explosive
proportions; reviews of this topic have appeared (Anonymous, 1977;
Bunn, Gabbay, and Gallop, 1978; McDonald *et al.*, 1978; Peterson and
Jones, 1977). Hb A_{Ia} has two glycophosphorylated components which
are not increased in diabetes. Although *in vitro* reaction of Hb A
with glucose-6-phosphate produces a hemoglobin with the chromatographic behavior of Hb A_{Ib}, Krishnamoorthy, Gacon, and Labie (1977)
as well as McDonald *et al.* (1978) report the absence of phosphate
in Hb A_{Ib}.

Hemoglobin A_{Id} probably is an artifact which forms *in vitro*
when undialyzed hemolysates are stored and in which one sulfhydryl
group of each of the two β chains of Hb A has reacted with oxidized
glutathione. This reaction has been studied in detail by Huisman
and Dozy (1962b), Huisman and Horton (1965), and Huisman *et al.*
(1966). Oxidized glutathione (GSSG) is formed in red-cell hemolysates upon aging and storage. The gradual conversion of glutathione (GSH) to GSSG will result in the formation of the Hb A_{Id}.

In the course of time, 50% of the available GSH will be bound by
hemoglobin molecules, but a molar ratio of GSH:Hb of 4:1 is required
for a complete conversion of normal Hb A into the complex. Assuming
normal GSH levels of the adult red blood cell hemolysates (GSH:Hb
ratio 0.3-0.6:1), it may be expected that ultimately 7.5-15% of the
Hb A will be transformed into the mixed disulfide. Such increases
in the Hb A$_I$ fractions have been observed in undialyzed hemolysates
which were stored for 2-3 months (Huisman *et al.*, 1966).

The effect of incubation of red cells at 37°C or of addition
of the Hb A-GSSG complex is also shown in Figure 8.1. Hb A$_{Ie}$ apparently is formed as a result of the incubation at 37°C, but its

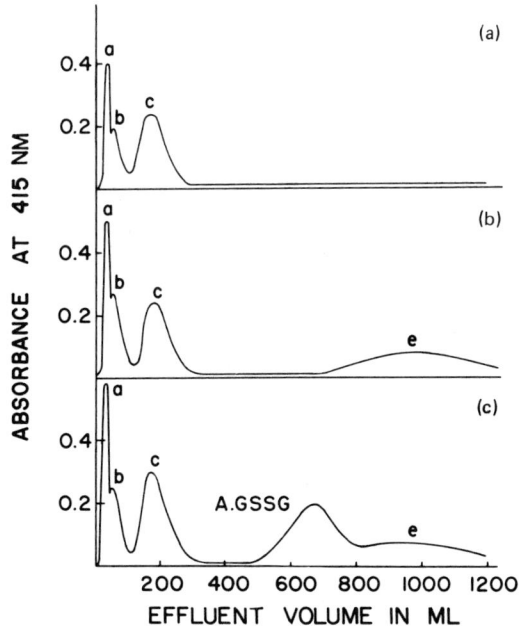

Figure 8.1. Amberlite IRC-50 chromatography of the minor hemoglobin components of adult red cell hemolysates under different experimental conditions. (a) Normal; (b) Normal, incubated for 24 hr at
37°C; (c) Same as (b), with small amount of A-GSSG added. (From
Huisman and Horton, 1965, with permission.)

chromatographic property is different from that of the Hb A-GSSG complex which presumably is the same as Hb A_{Id}.

When the isolated hemoglobins are studied by starch-gel electrophoresis a considerable heterogeneity is observed. As seen in Figure 8.2, Hb A_{Ic} is composed of one major fraction with an electrophoretic mobility slightly faster than that of Hb A and a minor fraction with an electrophoretic mobility similar to that of Hb F. Two fast-moving fractions are present in Hb A_{Ia}. Heterogeneity is also observed for Hb A_{Ib}. Because all isolated minor fractions consistently show fast moving "tails," heterogeneity is probably due to handling.

Hemoglobin A_{Ie} has not been characterized.

With the interest in the relationship of Hb A_{Ic} and diabetes has come the need for rapid methods for the quantitation of Hb A_{Ic}. As a result, the original methods of Allen, Schroeder, and Balog (1958) have been modified for use on a macro- and microchromatographic level. Some of these methods provide quantitative data both for Hb A_{Ia+b} and Hb A_{Ic} whereas others determine simply the sum. However, no method separates Hb A_{Ic} from Hb F. Consequently,

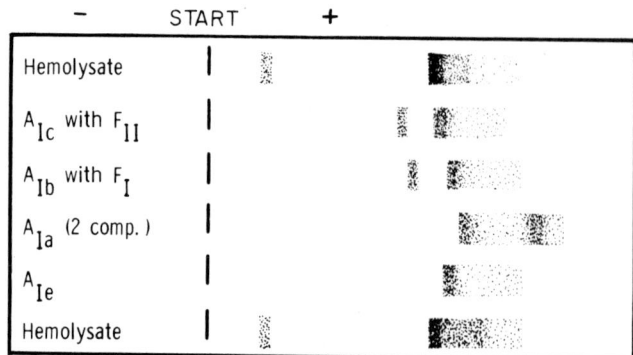

Figure 8.2. Tracing of the starch-gel electrophoretic patterns of several minor hemoglobin components isolated by Amberlite IRC-50 chromatography. (Modified from Huisman and Horton, 1965, with permission.)

an individual with an elevated Hb F might falsely be judged to have elevated Hb A$_{Ic}$ and diabetes.

II. MACROCHROMATOGRAPHIC PROCEDURES

A. Method of Allen, Schroeder, and Balog (1958)

This method has been discussed in detail in Chapter 5, Section II and needs little further comment. When Developer 5 (Table 5.1) is used, the separation of Hb A$_{Ia+b}$ and Hb A$_{Ic}$ is typified by Figure 5.3. According to Clegg and Schroeder (1959) who applied this procedure, normal individuals averaged 2.0% Hb A$_{Ia+b}$ (range, 1.7-2.4) and 5.7% Hb A$_{Ic}$ (range, 5.2-6.3). Approximately the same results came from the work of Horton and Huisman (1965) and Huisman and Horton (1965).

B. Method of Trivelli, Ranney, and Lai (1971)

1. *Procedure*

a. Developers

Two developers are used: Developer 6 (Table 5.1) and an undesignated high-phosphate developer which contains 6.52 g anhydrous Na$_2$HPO$_4$ and 14.35 g NaHPO$_4$·H$_2$O liter^{-1} at a pH of 6.42.

b. Pouring and Equilibration of Column

The 2 X 17.5 cm columns are poured from Bio-Rex 70 (the equivalent of Amberlite IRC-50, see comment on p. 23) in Developer 6 and equilibrated at room temperature with this developer for several days prior to initial use.

c. Development of the Chromatogram

The sample is 200-300 mg hemoglobin in 2-3 ml that had been equilibrated overnight at 4°C against Developer 6. After the top of the column has been stirred and resettled, the sample is applied and developed at the rate of 75-85 ml hr^{-1} with Developer 6 for 3 hr at room temperature. Hb A_{Ia+b+c} elutes with 225-250 ml in this period of time; although the separation of Hb A_{Ia+b} and Hb A_{Ic} is visually apparent, all are collected together. The remaining hemoglobin is eluted with the high-phosphate developer. Spectrophotometric readings are made at 552 nm.

    d. Reequilibration and Reuse

    Between runs, the column is equilibrated with Developer 6 for "at least 6 hr" — presumably at the rate of 75-85 ml hr^{-1}.

2. Application

Although large samples and columns are used in this method, it is reported that 10 determinations may be done simultaneously. Trivelli, Ranney, and Lai (1971) applied the procedure to 20 controls, 25 juvenile diabetics, and 75 adult diabetics. In the controls, the Hb A_I (that is, Hb A_{Ia+b+c}) was $6.5 \pm 1.5\%$ (SD) whereas the juvenile diabetics had $12.4 \pm 3.1\%$ and the adult diabetics $11.1 \pm 2.9\%$. There was considerable overlap of the ranges and many correlations were attempted.

C. Method of Fitzgibbons, Koler,
    and Jones (1976)

1. Procedure

This procedure is essentially a smaller scale modification of the method of Trivelli, Ranney, and Lai. The same developers are used, the column is 1 X 30 cm, the sample is 25-50 mg, the flow rate is 20-40 ml hr^{-1}, and 500 ml are used for reequilibration. Hb A_{Ia+b} elutes in the first 35 ml of effluent, Hb A_{Ic} in the next 100-150 ml, and Hb A_{II} in 100-200 ml of the high-phosphate developer.

2. Application

This method was used to measure the percentages of Hb A_{Ia+b} and Hb A_{Ic} in the 10% youngest and 10% oldest erythrocytes of normal and diabetic subjects after separation by centrifugation. Both cell age and diabetes are significant in determining the amount of Hb A_{Ia+b} and Hb A_{Ic}, as presented in Table 8.1.

Table 8.1 Hb A$_{Ia+b}$ and Hb A$_{Ic}$ as a Function of Cell Age in Normal and Diabetic Subjects

Subject	Percentage			
	Hb A$_{Ia+b}$		Hb A$_{Ic}$	
	Young cells	Old cells	Young cells	Old cells
Normal	1.2 ± 0.2	1.8 ± 0.4	3.1 ± 0.8	6.0 ± 1.1
Diabetic	1.7 ± 0.6	2.6 ± 0.9	5.1 ± 2.1	10.1 ± 3.7

D. Method of Kynoch and Lehmann (1977)

1. *Procedure*

  a. Developers

  Buffer No. 1 at pH 6.8 contains 5.12 g $NaH_2PO_4 \cdot H_2O$, 1.8 g Na_2HPO_4, and 0.65 g KCN liter^{-1}. For Buffer No. 2, the quantities are 14.35 g $NaH_2PO_4 \cdot H_2O$ and 6.25 g Na_2HPO_4 liter^{-1}.

  b. Pouring and Equilibration
    of the Column

  A 2 X 5 cm column is poured in a 20-ml syringe barrel and equilibrated with 50 ml Buffer No. 1.

  c. Development of the Chromatogram

  The sample is prepared from the washed cells of 0.5 ml anti-coagulated blood to which are added first 0.8 ml hemolyzing reagent (0.1 g powdered white saponin, and 0.5 g KCN/100 ml) for 1 min and then 9 ml Buffer No. 1. After centrifuging for 5 min "at 3000 rpm," the clear solution is placed on the column and developed with Buffer No. 1 at a flow rate of 150 ml hr^{-1} (liquid head: 40-50 cm). Hb A$_I$ is eluted in the first 100 ml and the remainder with 100 ml Buffer No. 2.

  d. Reequilibration and Reuse

  The column may be reequilibrated with 100-150 ml Buffer No. 1.

2. Application

It is stated that the values of 6-9% in nondiabetics and 6-19% in diabetics agree well with reported values.

E. Method of Abraham *et al.* (1978)

This method is essentially that of Trivelli, Ranney, and Lai (1971) and Fitzgibbons, Koler, and Jones (1976) with a return to the slow flow rates of Allen, Schroeder, and Balog (1958).

1. Procedure

As in the procedure of Fitzgibbons, Koler, and Jones (1976), a 1 X 30 cm column of IRC-50 (or Bio-Rex 70) is equilibrated with Developer 6 (Table 5.1) at room temperature. The sample is 25-30 mg hemoglobin that has been dialyzed against half-strength Developer 6. For development at room temperature, 225-300 ml Developer 6 are used at a flow rate of 14-15 ml hr^{-1} (16-20 hr) to elute Hb A_{Ia+b} and Hb A_{Ic}. Change to the high-phosphate buffer of Trivelli, Ranney, and Lai (1971) removes the remaining hemoglobin in 120-150 ml (8-10 hr). Throughout the chromatogram, 4.5-5.0 ml fractions are collected.

2. Application

The separations as depicted in Figure 8.3 are typical of those that Allen, Schroeder, and Balog (1958) obtained at 6-7°C with Developer 4 (Figure 5.3). The percentages as given in Figure 8.3 are well matched to the many that are reported for normal and diabetic subjects.

This procedure can be adapted for a study of the minor hemoglobins of other types of red blood cells. In unpublished work from the Medical College of Georgia, the minor hemoglobins of patients with sickle cell anemia (the so-called S_I components) were separated by applying a sodium phosphate gradient. The 1 X 30 cm column is equilibrated with Developer 6 which also fills a 500-ml constant-volume mixer. The influent is either Developer 7 (Developer 6 with 1.5 times the sodium phosphate concentration; 6.89 g $NaH_2PO_4 \cdot H_2O$, 1.77 g Na_2HPO_4, and 0.65 g KCN $liter^{-1}$) or Developer 8

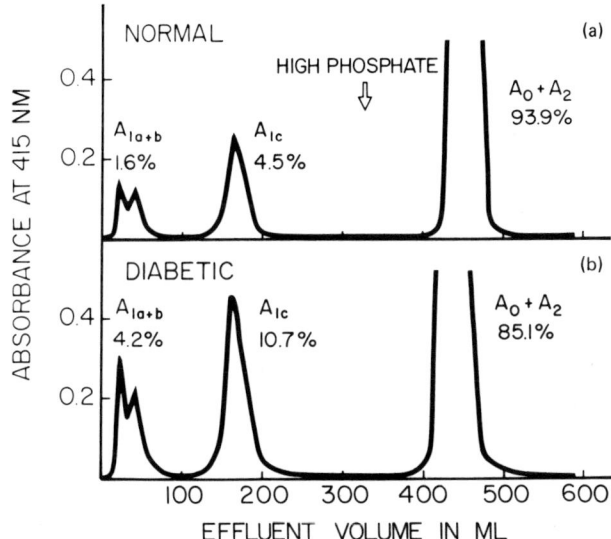

Figure 8.3. Chromatographic separation on 1 X 25 cm columns of Amberlite IRC-50 of the hemoglobins in a hemolysate from a normal control (a), and in a hemolysate from a diabetic patient (b) by the method of Abraham *et al.* (1978). (From Abraham *et al.*, 1978, with permission.)

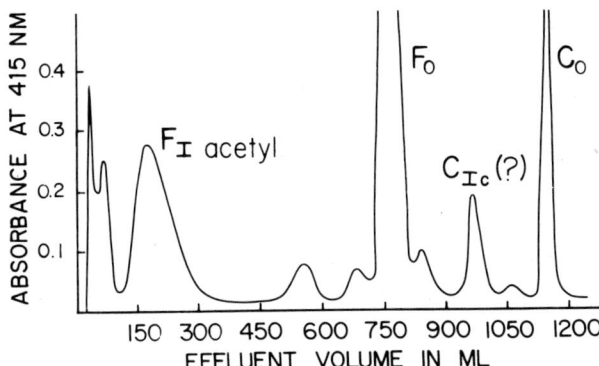

Figure 8.4. Chromatographic separation on a 1 X 25 cm column of Amberlite IRC-50 of the hemoglobin of a newborn baby with a homozygosity for Hb C. For further details, see text.

(Developer 6 with twice the sodium phosphate concentration; 9.18 g $NaH_2PO_4 \cdot H_2O$, 2.36 g Na_2HPO_4, and 0.65 g KCN $liter^{-1}$). The minor Hb F_I plus Hb F are eluted in the first 60 ml of effluent and are followed by several (often four) minor Hb S_I zones. Elution of the major Hb S zone (and of Hb A_2) has used a stronger developer (14.35 g $NaH_2PO_4 \cdot H_2O$, 6.52 g Na_2HPO_4, and 0.65 g KCN $liter^{-1}$). This is the strong developer of Trivelli, Ranney, and Lai (1971) (Section II.B) with KCN added.

The minor fetal hemoglobins in red cells from newborns can best be separated from Hb F_o by a gradient through a 1 X 30 cm column that has been equilibrated with half-strength Developer 6 (2.30 g $NaH_2PO_4 \cdot H_2O$, 0.59 g Na_2HPO_4, and 0.33 g KCN $liter^{-1}$). The 500-ml constant-volume mixer has this same buffer and the influent to the mixer is normal strength Developer 6 (Table 5.1). An example of the separations is shown in Figure 8.4 where the sample was the hemoglobin of a newborn with a homozygosity for Hb C. The most significant minor hemoglobin, Hb F_I, is readily separated from the other hemoglobin zones.

III. MICROCHROMATOGRAPHIC PROCEDURES

A. Method of Davis, McDonald, and
    Jarett (1977)

The method of Davis, McDonald, and Jarett (1977) uses high performance liquid chromatography (HPLC) with a column that is midway between macro- and microchromatographic size.

1. Procedure

    a. Developers

Developer 6 (Table 5.1) and a "high-phosphate buffer, pH 6.42" are used. The latter is not further characterized. Probably, it is similar to or identical with the strong developer of Trivelli, Ranney, and Lai (1971) which surely would be adequate.

   b.   Column

   The 0.9 X 7 cm column is poured from Bio-Rex 70 (-400 mesh)
and is equilibrated with Developer 6.

   c.   Development of the Chromatogram

   A 3-μl sample of hemolysate from washed erythrocytes (concentration of solution unspecified) is applied and developed with
Developer 6 at a flow rate of 1.3 ml min^{-1}. Hb A$_{Ia+b}$ and Hb A$_{Ic}$
emerge as separate peaks in 10 min and the remaining hemoglobin
with the high-phosphate buffer in an additional 10 min. Absorbance
was monitored at 410 nm and the percentages were calculated from
the areas.

   d.   Reequilibration and Reuse

   The column is reequilibrated with Developer 6 for 5 min.   It
should be noted that, if the flow rate during this equilibration is
1.3 ml min^{-1}, this quantity seems inadequate for reequilibration.

2. Application

The above description is based on an abstract which leaves some
details incomplete. It is reported that the Hb A$_{Ic}$ was 6.4 ± 0.6%
(SD) in 9 nondiabetic patients and 11.6 ± 2.4% in 13 juvenile diabetics who required insulin.

B. Method of Cole *et al.* (1978)

This is another HPLC method with a somewhat smaller column than
that of Davis, McDonald, and Jarett (1977) (Section A above).

1. Procedure

   a.   Developers

   Developer 6A has 1.65 times the concentration of Developer 6
(Table 5.1) and contains 7.58 g NaH$_2$PO$_4$·H$_2$O, 1.96 g Na$_2$HPO$_4$, and
1.07 g KCN liter^{-1}. The high-phosphate buffer has twice the con-

centration of the strong developer of Trivelli, Ranney, and Lai (1971) and contains 28.70 g $NaH_2PO_4 \cdot H_2O$ and 13.04 g Na_2HPO_4 liter^{-1}.

   b. Column

   The 0.9 X 5.8 cm column of Bio-Rex 70 (37-44 μ) was poured in high-phosphate buffer, inverted, and pumped at 23 ml hr^{-1} for 30 min; this is the "standard" column. A column for "rapid" assay is only 0.9 X 2.8 cm.

   c. Development of the Chromatogram

   Hemolysate is prepared by standard methods and the hemoglobin is converted to the carbonmonoxy form. For the final sample, 0.05 ml carbonmonoxyhemoglobin solution is diluted with 3 ml ice-cold Developer 6A. The 10-μl sample of this dilution which is injected onto the column contains about 12 μg carbonmonoxyhemoglobin.

   Each day for the standard procedure, the column is inverted, washed with high-phosphate buffer for 30 min at 23 ml hr^{-1} and then with Developer 6A at the same rate. Two blank chromatograms are run to provide corrections for absorbance change with buffer change. After injection of the sample, Developer 6A is used for 10 min, high-phosphate buffer for 4 min, and Developer 6A for 13 min for reequilibration. The progress of the chromatogram is followed by recording the output from a dual wavelength detector which uses wavelengths of 405 and 546 nm at full scale sensitivity of 0.01 and 0.08 absorbance units, respectively. Calculation is done by compensating planimetry or with a computing integrator.

   The rapid assay uses the shorter column with a flow rate of 40 ml hr^{-1}. The cycle after injection of sample is 3 min of Developer 6A, 2 min of high-phosphate buffer, and 6 min of reequilibration with Developer 2A.

2. Application

Figure 8.5 is an example of the chromatographic separations that are achieved by the standard procedure. The separation of Hb A_{Ia+b} from Hb A_{Ic} is incomplete but is not considered to introduce error

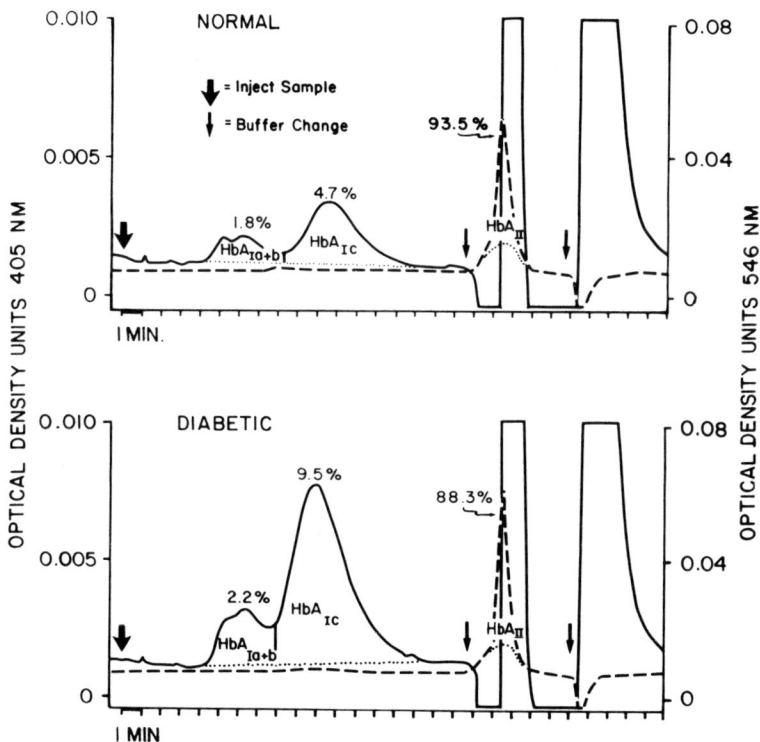

Figure 8.5. Chromatograms of hemoglobin of normal and diabetic individuals by the HPLC method of Cole *et al.* (1978). Broken lines = optical density units of 546 nm; solid line = optical density units of 405 nm. (With permission.)

into the quantitative determination. By the rapid method, only the sum of Hb A_{Ia+b+c} is determined. Reproducibility of the method was examined in several ways. A total of 36 determinations on 14 samples from a normal individual over a period of one month by the standard method gave Hb A_{Ic} of 4.95 ± 0.12% (SD) ± 0.2% (SEM) with CV = 2.4% and Hb A_{Ia+b} of 1.65 ± 0.06% ± 0.01% (CV = 3.6%); the Hb A_{Ia+b+c} by the rapid method was 6.61 ± 0.14% ± 0.02% (CV = 2.1%). Somewhat similar statistics resulted when a hemolysate was stored at 4°C and analyzed 15 times in 7 days. Hb A_{Ic} in 10 normal individuals ranged from 4.6 to 5.6% with a mean of 5.06 ± 0.32% ± 0.10%, whereas their Hb A_{Ia+b} was 1.68 ± 0.11% ± 0.03%. In 15 diabetic individuals, Hb A_{Ic} ranged from 6.8 to 20% and Hb A_{Ia+b} from 2.0 to 3.7%.

C. Method of Jones, Koler, and
    Jones (1978)

This method is a microchromatographic adaptation of the method of Trivelli, Ranney, and Lai (1971) and uses the basic microchromato-graphic techniques that were described in Chapter 7.

1. Procedure

   a. Developers

Developer 6 (Table 5.1) and the high-phosphate developer of Trivelli, Ranney, and Lai (see Section II.B above) are used.

   b. Pouring and Equilibration
       of the Column

A 9-10 cm column of Amberlite IRC-50 which has been equili-brated with Developer 6 is poured with a 1:3 suspension in a Pasteur pipet and equilibrated at room temperature with at least 40 ml Developer 6 at a flow rate of 3-4 ml hr^{-1}. After equilibration, the length is about 8 cm.

   c. Development of the Chromatogram

The sample which is prepared from 3 drops of hemolysate (10-12 g dl^{-1}) and 9 drops of Developer 6 is applied to the column, layered over with Developer 6, and attached to a funnel with a supply of Developer 6 so that the liquid head is about 20 cm. The effluent is collected in three fractions whose limits are determined visually. Hb A_{Ia+b} is eluted in the first 2.3-3.3 ml effluent, at which time the zone of Hb A_{Ic} is approaching the bottom of the column (at least 3/4 of the way down the column). The elution of Hb A_{Ia+b} is com-plete when the developer in the stem of the column and the drops of effluent are colorless against a white background. At this time (25-35 min from the start), collection of Hb A_{Ic} is begun and com-pleted in the next 4.5-6.5 ml of effluent (about 90 min). At that time, the major Hb A$_{II}$ is about half-way down the column. When the lower part of the column and developer in the stem are colorless, remaining Developer 6 is removed from the column and replaced with the high-phosphate buffer to remove Hb A$_{II}$. In order to speed this step, a meter length of tubing (1/32 cm inner diameter) is attached to the stem of the column and the effluent is collected in a volumetric flask which is placed on the floor. The major fraction is removed within 30 min. Spectrophotometric determination is made at 415 nm.

    d. Reequilibration and Reuse

    The columns may be reequilibrated by passing through at least 40 ml Developer 6. If, with continued use, the flow rate decreases to less than 2.5 ml hr^{-1}, the column should be discarded.

2. Application

Jones, Koler, and Jones (1978) applied this microchromatographic procedure and the method of Trivelli, Ranney, and Lai (1971) to the same diabetic and nondiabetic subjects. The agreement of results by the two methods was excellent as depicted in Figure 8.6.

    This procedure is rapid and applicable to multiple simultaneous determinations.

D. Method of Abraham *et al.* (1978)

This method measures only the quantity of Hb A$_I$ (= Hb A$_{Ia+b+c}$). The columns which are about 6 cm long are equilibrated with Developer 6. For the sample, 1 drop of blood is mixed with 8 drops of water. After hemolysis for 5 min, 5 drops are applied. Hb A$_I$ is eluted with 4 ml Developer 6 and the remaining hemoglobin with the high-phosphate buffer of Trivelli, Ranney, and Lai (1971) (Section II.B above). Spectrophotometric readings at 415 nm are the basis for the calculation of the percentage Hb A$_I$.

    Among other observations, they noted that storage of blood (not hemolysate) at 4°C for a month gave fairly constant values of Hb A$_I$.

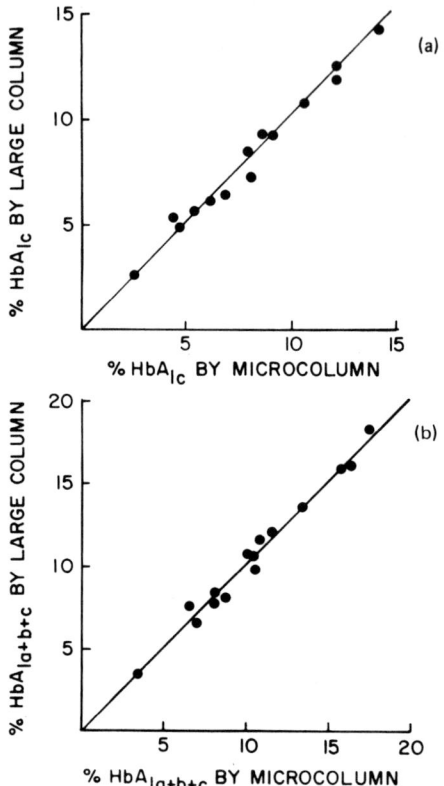

Figure 8.6. Correlation between the amounts of (a) Hb A_{Ic} and (b) Hb A_{Ia+b+c} as determined by the macrochromatographic method of Trivelli, Ranney, and Lai (1971) and the microchromatographic method of Jones, Koler, and Jones (1978). For (a), $y = 1.0x - 0.10$, $R^2 = 0.98$; for (b), $y = 0.96x + 0.13$, $R^2 = 0.98$. (From Jones, Koler, and Jones, 1978, with permission.)

A kit which uses this procedure is available from Isolab, Inc., Akron, Ohio, with instructions for use.

IV. COMMENTS

The several procedures that have been described above provide varied and both simple and more complex means for the quantitative determination of Hb A_{Ic}. Consequently, those who are interested in

examining the relationship of Hb A$_{Ic}$ to diabetes have at their dis-
posal rapid and accurate procedures for the determination of Hb A$_{Ic}$.

All of the procedures are variations on the methods of Allen,
Schroeder, and Balog (1958) and all appear to have comparable accuracy and precision. Thus, data for normal individuals as reported
by Trivelli, Ranney, and Lai (1971) accord with those of Clegg and
Schroeder (1959) who used Allen's method. The microchromatographic
procedure of Jones, Koler, and Jones (1978) was shown by these authors to provide data that are comparable with those from the Trivelli
method. Abraham *et al.* (1978) compared their macrochromatographic
procedure (Section II.E above) with their modification of the microchromatographic modification (Section III.D) as represented by the
commercially available Isolab kits. The data are shown in Figure 8.7;

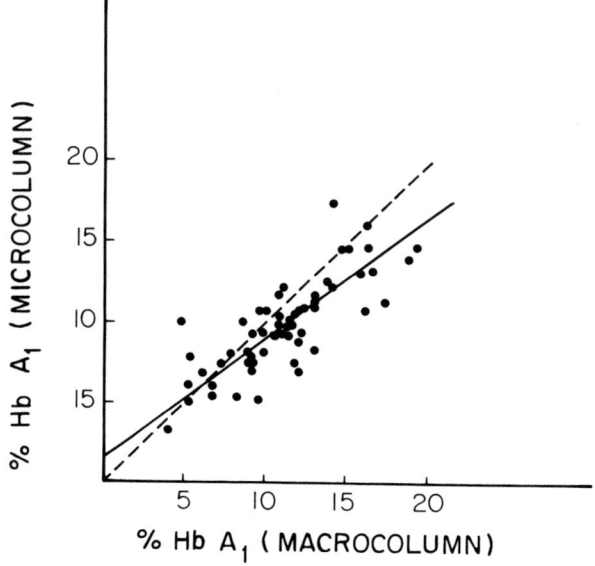

Figure 8.7. The correlation of the Hb A$_I$ values as determined by
microchromatography (Isolab columns) and macrochromatography (Section II.D). The solid line is the least-squares regression line
and the broken line is the theoretical line of correlation. n = 60;
R = 0.86; and y = 0.75x + 1.6. (From Abraham *et al.*, 1978, with
permission.)

the microchromatographic procedure tended to be slightly lower. Their studies provided a comparison with the colorimetric method for Hb A_{Ic} of Flückiger and Winterhalter (1976). Their results when various methods were applied to patients with different ratings of diabetic management are presented in Figure 8.8. The correlation of Hb A_I and A_{Ic} levels against the different management groups is similar for both column chromatographic procedures. Correlation between management groups and Hb A_{Ic} levels determined by the colorimetric procedure, however, is decidedly poorer. Their data support

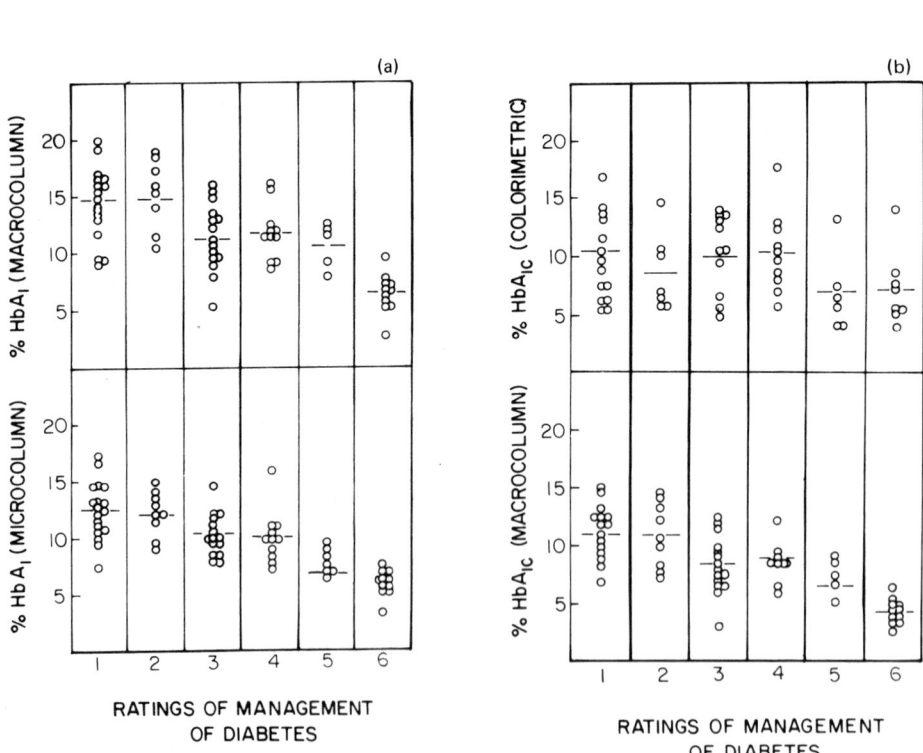

Figure 8.8. (a) The levels of Hb A_I in patients with juvenile diabetes who had varying degrees of diabetic management. (b) The levels of Hb A_{Ic} in juvenile diabetics. Comparison of the macrocolumn procedure of Trivelli, Ranney, and Lai (1971) and the colorimetric procedure of Flückiger and Winterhalter (1976). (From Abraham *et al.,* 1978, with permission.)

the conclusion above that many reliable methods for the determination of Hb A$_{Ic}$ are now available.

The initial observations of Trivelli, Ranney, and Lai (1971) that the percentage of Hb A$_{Ia+b+c}$ in diabetics is approximately double that in normal individuals and the findings of Koenig *et al.* (1976) that the percentage of Hb A$_{Ia+b}$ and Hb A$_{Ic}$ return to normal levels in diabetics under "control" have been repeatedly confirmed (see brief reviews by Anonymous, 1977; Bunn, Gabbay, and Gallop, 1978; Cerami, Koenig, and Peterson, 1978; and Peterson and Jones, 1977, for references). The many chromatographic methods now available allow the interested investigator to choose whether he wishes to determine Hb A$_I$ or Hb A$_{Ia+b}$ and Hb A$_{Ic}$.

It has already been mentioned and should be emphasized that no procedure that is described in this chapter will separate Hb A$_{Ic}$ from Hb F. If, therefore, such methods were applied to an individual with an elevated percentage of Hb F, a false interpretation is likely. It may be that the differential examination of Hb A$_{Ia+b}$ and Hb A$_{Ic}$ would distinguish between increased Hb A$_{Ic}$ and Hb A$_{Ic}$ plus Hb F; thus, a normal percentage of Hb A$_{Ia+b}$ and an increased percentage of "Hb A$_{Ic}$" might suggest normal Hb A$_{Ic}$ and elevated Hb F. Further means of distinguishing the presence of an increased level of Hb F would be the well-known alkali denaturation methods. Chromatographically, the methods of Chapter 9 for Hb F would also permit this distinction.

CHROMATOGRAPHY OF HEMOGLOBIN F AND HEMOGLOBIN γ_4

I. INTRODUCTION

Normally, the Hb F of the newborn infant decreases rapidly in
amount and, by the age of 6 months to 1 year, accounts for at most
a few percent of the total hemoglobin. The normal adult will have
Hb F of the order of 0.5 ± 0.5%, although the occasional hematolog-
ically normal adult will have perhaps as much as 5% Hb F. However,

148

Hb F is elevated in percentage in a wide variety of hematological
conditions. Some states in which Hb F is elevated are benign in
their expression. For example, heterozygotes for the hereditary
persistence of fetal hemoglobin (HPFH) may have 5-30% Hb F and
homozygotes for most types of HPFH produce only Hb F. Both heterozygotes and homozygotes for HPFH are in good health and show minimal and unimportant hematological differences (if any) from normal.
On the other hand, Hb F is also elevated in other conditions in
which the individual is in poor health. In such instances, the
elevated percentage of Hb F appears to be compensatory, perhaps a
reversion to the fetal condition in order to provide a working hemoglobin to permit the individual to survive. For example, the homozygote for β^0-thalassemia will have only Hb F and a few percent of
Hb A_2. Although, on the basis of hemoglobin composition, this individual differs little from the HPFH homozygote, the β^0-thalassemia
homozygote is severely affected because of his anemia. In any one
disease state, Hb F may be present in rather variable amount. Individuals with sickle cell anemia may vary as much as an order of
magnitude, say from a few percent to thirty percent. Elevated
levels of Hb F occur in various leukemias, anemias, and miscella-
neous conditions in varied amounts and with varied frequency. This
general topic of Hb F in abnormal conditions has been reviewed by
Weatherall, Pembrey, and Pritchard (1974).

Because of the increase of Hb F in many abnormal conditions,
its quantitative determination has been of considerable interest
and use in diagnosis. The resistance of Hb F to alkali denaturation in contrast to the lability of Hb A has led to many descriptions of alkali denaturation procedures that are based on these
properties. Although such methods have received widespread use,
they appear usually to overestimate the amount of Hb F when the
percentage is low and vice versa. Most chromatographic procedures
on either cation or anion exchangers (see Chapters 5 and 6) provide
an overestimate of Hb F because they commonly fail to separate Hb F
and Hb A_1 (or on occasion, even Hb F and Hb A). If Hb A and electrophoretically fast moving hemoglobins are absent however, they

provide an accurate estimate of Hb F.   In an attempt to determine
Hb F more accurately in the presence of Hb A, a procedure which is
based on an amino acid analysis of Hb F + A_1 from diethylaminoethyl-
(DEAE)-Sephadex chromatography was devised (Schroeder *et al.*, 1970).
Although this method appears to be accurate, it is cumbersome and
time-consuming.

Although most chromatographic procedures fail to separate Hb F
and Hb A_1 (or Hb A_{Ic}), it is now possible to do so by chromatography
on DEAE-cellulose under appropriate conditions.   This chapter describes and evaluates such procedures.

Hemoglobin γ_4 or Bart's is detectable at birth in some cord
bloods.   Its presence is generally considered to be indicative of
an α-thalassemia determinant and its quantity to define the specific α-thalassemia condition.   This topic is discussed in detail by
Weatherall and Clegg (1972).   In the concluding section of this
chapter, both macro- and microchromatographic means of determining
Hb γ_4 are described.

II.   MACROCHROMATOGRAPHIC DETERMINATION
   OF Hb F

A.   Procedure

The procedure has been described in detail in Chapter 6, Section
III.A.   In summary, about 30 mg dialyzed hemolysate (against Developer A) is applied to a 1 X 25 cm column of DE-52 which has been
equilibrated with Developer A at pH 7.8 ± 0.05.   Development is by
gradient from a 500-ml constant-volume mixer which initially contains Developer A-5 and into which Developer A-20 is introduced.
The flow rate is 15-18 ml hr^{-1} and 5-6 ml fractions are collected.
Ad lib after about 24 hr, Developer A-20 is replaced by Developer
A-40, and the chromatogram is continued for another day.

B.   Examples

Figure 9.1a typifies the separation of the components of adult hemoglobin under the above conditions.   No Hb F zone as such is apparent.

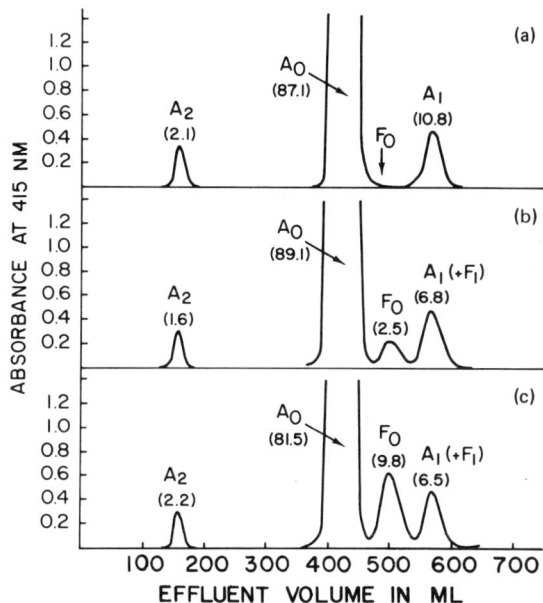

Figure 9.1. DEAE-cellulose chromatograms of red-cell hemolysates
with different quantities of Hb F. (a) Normal adult. (b) and (c)
Mixtures of hemolysates of a normal adult and a cord blood. The
numbers give the percentages of the hemoglobins. Fraction size,
5-6 ml. (From Abraham *et al.*, 1977b, with permission.)

However, when some cord blood is added, a well-defined zone of Hb F
appears between the well-separated zones of Hb A and Hb A_1. The
method is sensitive because a pronounced peak of Hb F is evident
even at 2.5% (Figure 9.1b). It is somewhat surprising that, in
this example, the sum of Hb A_1 + F_1 in the mixtures is less than
that of Hb A_1 in the adult sample.

The utility of the method in separating Hb F in a variety of
conditions with hemoglobin variants is demonstrated in Figures 9.2
and 9.3. The Hb F is easily separated from complex mixtures, and
the percentage not only of it but of the other hemoglobins may be
calculated without difficulty.

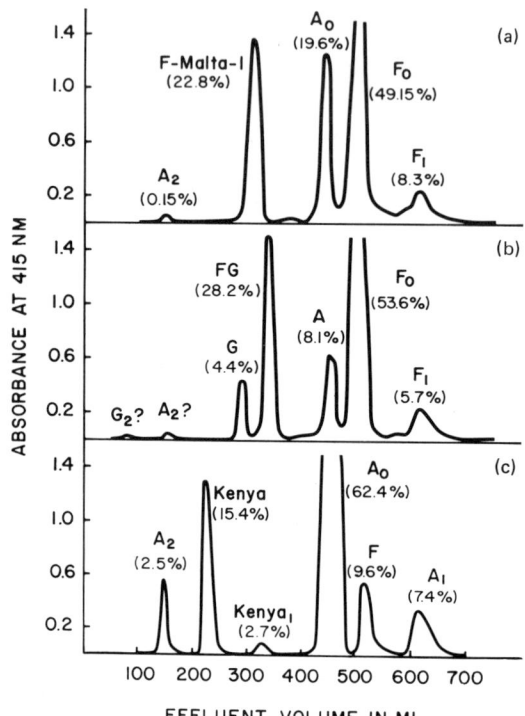

Figure 9.2. Chromatographic separations on columns of DEAE-cellu-
lose of the hemoglobins of a newborn with the γ-chain variant Hb
F-Malta I (a), those of a newborn with the α-chain variant Hb G-
Philadelphia (b), and those of an adult with Hb Kenya trait (c).

C. Quantitative Determination of Hb F

In assessing the reliability of this macrochromatographic procedure
for Hb F, Abraham *et al.* (1977b) examined test solutions and conditions and then compared with the results of other methods for Hb F.
Thus, two mixtures of adult and cord hemoglobins were analyzed repeatedly over a period of 10 days. The percentage of Hb F from six
determinations on one sample ranged from 11.2 to 12.7 with a mean of
12.1 and SD of 0.6; for the other sample, the data were 24.5-27.4
with a mean of 26.2 and SD of 1.2.

      In other experiments, mixtures of blood from a normal adult
and from an HPFH homozygote with 100% Hb F were examined by four

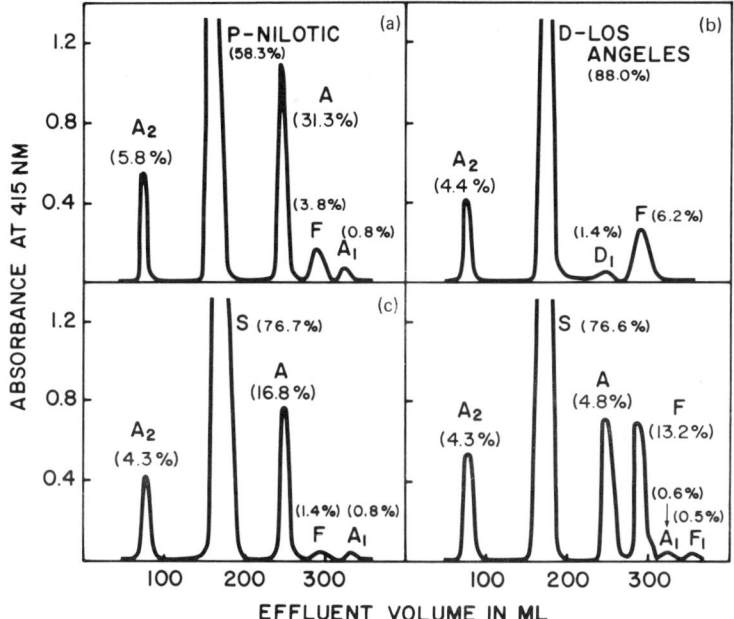

Figure 9.3. Chromatographic separation on columns of DEAE-cellulose of the hemoglobins of persons with a combination of β-thalassemia and either Hb P-Nilotic (a), Hb D-Los Angeles (b), or Hb S (c and d).

different methods for determining Hb F; these procedures were the present one (F_{DE}), an alkali denaturation method (F_{AD}) (Betke, Marti, and Schlicht, 1959), a chemical procedure (F_{Ile}) (Schroeder *et al.*, 1970), and an immunochemical method (F_{RIA}) (Garver *et al.*, 1976). The results of these comparisons are depicted in Figure 9.4.

D. Comments

The method gives an excellent separation of Hb F from Hb A and per-mits the quantitation of the former in the presence of Hb A. In chromatograms of hemoglobin from normal adults (Figure 9.1), no discrete zone of Hb F is visible. Consequently, it is not possible to measure Hb F accurately by this method if the quantity is less than about 2%. However, even an elevation of 2-3% will become apparent (Figure 9.1) and may be calculated although the result

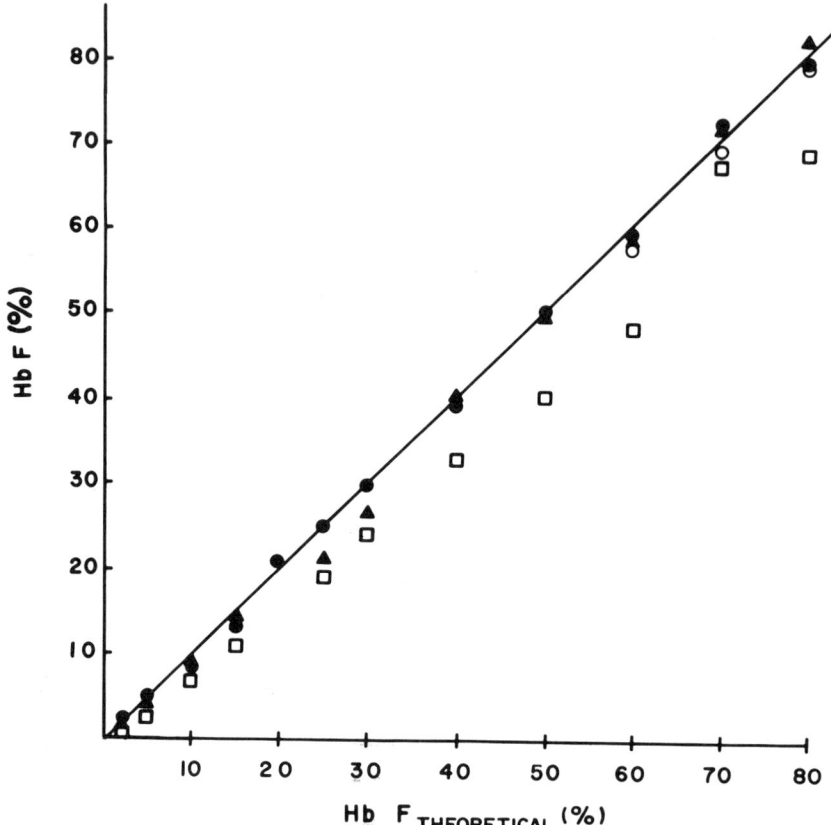

Figure 9.4. The percentages of Hb F in mixtures of hemolysates of a normal adult and an HPFH homozygote (with 100% Hb F). F_{DE} = % Hb F by DEAE-cellulose chromatography (solid circles); F_{Ile} = % Hb F by a chemical procedure (open circles); F_{AD} = % Hb F by alkali denaturation (open squares); F_{RIA} = % Hb F by radioimmunoassay (solid squares).

probably will be significantly in error because of the slight tailing of the Hb A.

As the amount of Hb F increases to 10% and above, Hb F_1 will begin to be present in significant amount. Hb F and Hb F_1 separate by this chromatographic method, but Hb F_1 usually coincides with Hb A_1. The results from this determination, therefore, do not include

Hb F_1 which apparently differs from Hb F mainly in the presence of N-acetylated γ chains (Schroeder *et al.*, 1962; Stegink, Meyer, and Brummel, 1971) and is a postsynthetic modification. It is somewhat surprising that % F_{DE} (this method) agrees so well with the results of % F_{Ile} and % F_{RIA} (Figure 9.4) both of which determine the sum of Hb F and Hb F_1. On the other hand, in samples from which Hb A is absent, the sum of Hb F and Hb F_1 provides a correct measure of the % Hb F.

III. MICROCHROMATOGRAPHIC DETERMINATION OF Hb F

The first procedure to be described is a microchromatographic adaptation of the method in the prior section but without gradient; it uses the drinking straws to hold the column (Chapter 7, Section VI). The second method resembles the microchromatographic method for hemoglobinopathy detection at birth (Chapter 7, Section III); it may be used only in the absence of Hb A.

A. General Determination of Hb F by the Drinking Straw Method

1. *Procedure*

The procedure follows exactly that in Chapter 7, Section VI. In summary, 3 drops of blood are hemolyzed in water and applied to a DE-52 column in a drinking straw. After the chromatogram has been developed with 60 ml Developer A-15 (0.2 M glycine/0.15 M NaCl/0.01% KCN), the zones are cut out, eluted, and examined spectrophotometrically.

2. *Examples*

Figure 7.8b is a drawing of a typical chromatogram at the end of development before the straw is cut and the zones are eluted. Hb F is not apparent as a discrete zone when a sample from a normal adult is chromatographed, but it becomes obvious if only 1-2% of cord blood is added. When the amount of Hb F in a sample is small, it is advisable to use a simultaneous control in which Hb F is definitely elevated. In this way, the positions at which the cuts should be made can be more precisely delineated.

3. *Quantitative Determination of Hb F*

In a series of 12 simultaneous determinations with the same sample of adult blood to which a small amount of cord blood had been added, the results by this method which are denoted "F_{St}" averaged 5.1% with a range from 4.8 to 5.5%. In the same series, Hb A_2 was 2.4% (range 2.3-2.5%), Hb A was 83.3% (range 82.5-83.8%), and Hb $A_1 + F_1$ was 9.2% (range 8.9-9.4%). In a large series of duplicate determinations, the two values with few exceptions agreed within 10% of each other despite wide variation in the percentage of Hb F.

When the method was applied to 219 adults with normal hematological indices, the results of Figure 9.5 were obtained. A slight

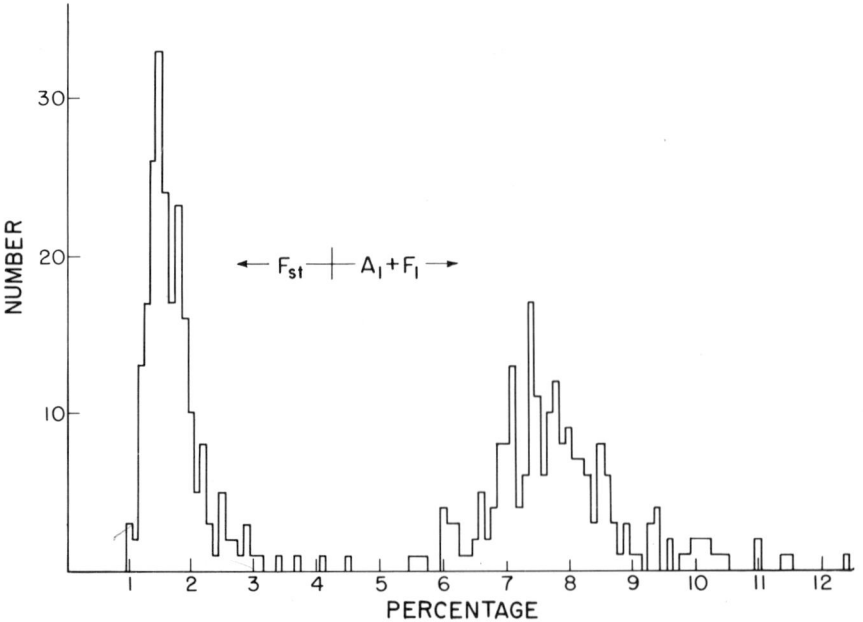

Figure 9.5. Range of values for % F_{St} and % Hb $A_1 + F_1$ in samples from normal adults. (From Schroeder, Pace, and Huisman, 1978, with permission.)

tailing of the Hb A is responsible for the values. The % F_{AD} (F by alkali denaturation) from 36 of 37 randomly selected samples of the series ranged from 0.4 to 1.6%; one sample had 3.0% F_{St} and 3.2% F_{AD}. Although the method provides an inexact measure of Hb F in the normal adult, the normal range is about 1-2.5% F_{St}. Thus, when Hb A is the major hemoglobin, the % F_{St} will be 1-2% higher than the actual Hb F because of the tailing of Hb A_o. The Hb A_1 + F_1 tends to fall between 6 and 9% (Figure 9.5). Other analyses of a variety of samples with elevated Hb F may be found in Table 7.3.

4. *Comments*

This micromethod has been compared with other procedures for Hb F (Schroeder, Pace, and Huisman, 1978). As anticipated, there is excellent agreement between % F_{St} and % F_{DE} (macromethod; Section II above). However, a concordance between % F_{St} and % F_{AD} is fortuitous; the % F_{St} does not include Hb F_1. If % F_{St} is determined in samples from sickle cell anemia patients in whom Hb A_1 is of course absent, Hb F_1 may also be estimated. The sum of % F_{St} + % F_1 in such patients is greater than % F_{AD}. The sum of % F_{St} + % F_1 by this method agrees well with % F_{micro} which is determined as described below and which determines Hb F and Hb F_1.

Although the presence of Hb A_1 in Hb A-containing samples prevents a direct determination of Hb F_1 and, therefore, a true measure of total Hb F, the sum of % F + F_1 may be estimated from % F_{St}. Thus, when fresh cord blood samples are chromatographed on the straws, the ratio of % F_1 + other components to % F_{St} is 1:4 to 1:3. This is comparable to previously reported data (Matsuda *et al.*, 1960). Likewise, in 18 SS samples with % F_{St} above 10%, the ratio of % F_1 to % F_{St} averaged 1:3. Because the quantity of Hb F_1 is 20-25% that of the F_{St} in cord-blood samples and SS patients, the total Hb F in an Hb A-containing sample may reasonably be calculated to be 1.25 F_{St}. Such a calculation may also be applied to values of % F_{DE} (Section II, above) which do not measure Hb F_1.

Although radioimmunoassays or radial immunodiffusion techniques are more difficult than chromatography, such methods as described by

Boyer *et al.* (1974), Chudwin and Rucknagel (1974), Garver *et al.*
(1976), and Shukla and Headings (1974) may provide a more accurate
way to determine Hb F quantitatively.

B. Determination in the Presence
    of Hb S and/or Hb C

The method was initially described by Schroeder *et al.* (1976).

1. Procedure

    a. Developers

    The developer is Developer C-1 as already described in Chapter
7, Section IV. It contains 0.03 M bis-Tris-HCl/0.04 M NaCl/0.01%
KCN at pH 6.2 and is prepared from 6.28 g bis-Tris, 2.34 g NaCl,
0.1 g KCN plus concentrated HCl to pH 6.2 in 1 liter.

    b. Preparation of the Ion
       Exchanger

    Microgranular, preswollen CM-cellulose (CM-52; Whatman, Clifton,
N.J.) is equilibrated with Developer C-1.

    c. Preparation of the Sample

    Samples may be prepared in several ways. If a conventional
hemolysate at 8-10% concentration is available, 0.02 ml is mixed
with 0.3 ml Developer C-1 and 0.2 ml 0.004 M maleic acid.

    If blood is used, 0.02 ml is mixed with 0.3 ml Developer C-1,
which contains 0.05% saponin, and with 0.2 ml 0.004 M maleic acid.

    If blood on filter paper is available, an area equivalent to
a circle about 1.0-1.5 cm in diameter (on Whatman 3MM filter paper
this equals 0.02-0.03 ml blood) is cut into small pieces and submerged in 0.3 ml Developer C-1 with saponin and 0.2 ml 0.004 M
maleic acid. After occasional stirring over 5-10 min, as much
fluid as possible is removed and used as the sample.

    d. Development of the Chromatogram

    A 0.5 X 6 cm column of CM-cellulose is poured in a Pasteur
pipet. After application of the sample, development is made with
Developer C-1. When blood is used as the sample, the first few

milliliters of effluent contain cell debris and should be discarded. The next 20 ml that contain the Hb F are then collected. Finally, the remaining hemoglobin is removed with 2% KCN. Spectrophotometric reading is made at 415 nm. If the liquid head above the column is 50-60 cm, the procedure can be completed in about 1½ hr.

2. *Quantitative Determination of Hb F*

In order to evaluate the precision of the method, 20 determinations per day were made on three successive days on the same sample of whole blood. The 60 values averaged 16.6% with a range from 15.7 to 17.1% and an SD of ± 0.3%. In other experiments, 10 determinations per day were done on samples from five different individuals with these results: 2.7 ± 0.3% (range 2.4-3.3), 5.9 ± 0.4% (5.2-6.8), 15.4 ± 0.4% (14.6-16.0), 19.9 ± 0.9% (18.5-21.5), and 33.9 ± 1.9% (30.7-36.6).

In the original study (Schroeder *et al.*, 1976), the results of this F_{micro} method were compared with other methods with results that are depicted in Figure 9.6. Not unexpectedly there is disagreement with F_{AD} measurements which tend toward lower values than chromatographic methods provide (see also Figure 9.4). Good correlation with F_{DES} is evident, but the F_{micro} method yields somewhat higher percentages than the F_{Ile} method. This is to be anticipated because F_{Ile} data usually are 85-90% of F_{DES} results. From these data and others with artificial mixtures of bloods from SS patients with very low percentages of Hb F and from homozygotes for HPFH or β^0-thalassemia, it was concluded that the F_{micro} method may provide somewhat high results at low percentage of Hb F and vice versa. It appears as though the accuracy and precision is ± 5-10%.

From experiments with storage of blood, it was concluded that blood may be stored at 4°C for as long as 30 days without unduly influencing the results.

3. *Comments*

It should be emphasized that this F_{micro} method is not applicable if Hb A is present as, for example, in a normal individual, an HPFH heterozygote, a sickle cell trait individual, etc. However, if only

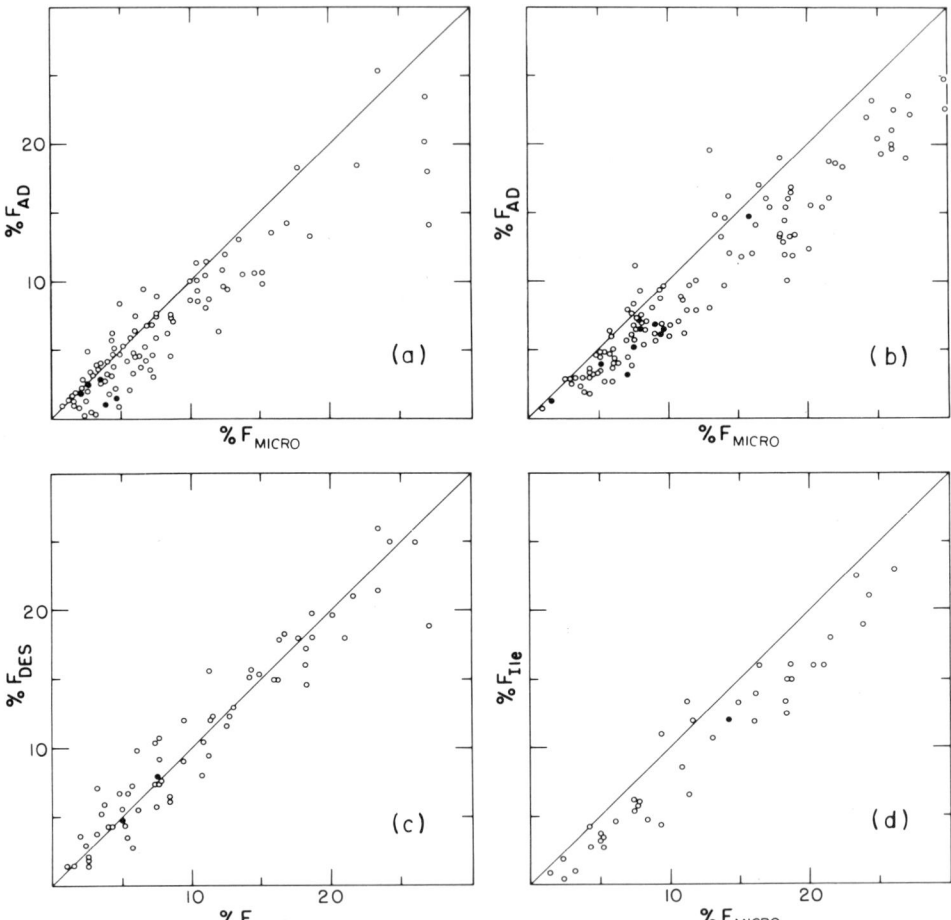

Figure 9.6. Percentage of F_micro as compared to results of other methods for Hb F on the same sample. Comparisons are as follows: In (a), with the alkali denaturation method of Betke, Marti, and Schlicht (1959) (F_{AD}); in (b), with a modified alkali denaturation method (Schroeder *et al.*, 1976); in (c), with conventional DEAE-Sephadex chromatography (Chapter 6, Section II) (F_{DES}); and in (d), with amino acid analysis of the DEAE-Sephadex peak (Schroeder *et al.*, 1970) (F_{Ile}). Filled circles represent more than one value. SS, SC, CC, and S-HPFH samples are included. (From Schroeder *et al.*, 1976, with permission.)

Hb S and/or Hb C (and probably also other variants with similar chromatographic properties) are present, it provides a rapid means for the quantitative determination of Hb F + Hb F_1.

IV. QUANTITATIVE DETERMINATION
     OF Hb γ_4

A. Macrochromatographic Method

This procedure is an adaptation of chromatography on CM-cellulose with bis-Tris-NaCl developers as described in Chapter 5, Section V. The conditions are similar to those in Figure 5.16 except that no gradient is used.

1. Procedure

   a. Developers

   The three developers are: Developer BT-4 which is 0.03 M bis-Tris/0.01% KCN (6.28 g bis-Tris and 0.1 g KCN liter^{-1}); Developer BT-5 which is 0.03 M bis-Tris/0.005 M NaCl/0.01% KCN (6.28 g bis-Tris, 0.29 g NaCl, and 0.1 g KCN liter^{-1}); and Developer BT-6 which is 0.03 M bis-Tris/0.1 M NaCl/0.01% KCN (6.28 g bis-Tris, 5.83 g NaCl, and 0.1 g KCN liter^{-1}).

   b. Preparation of the Ion
       Exchanger

   Microgranular, preswollen CM-cellulose (CM-52; Whatman, Clifton, N.J.) is equilibrated with Developer BT-4 at pH 6.1.

   c. Pouring and Equilibration
       of the Column

   A 1 X 30 cm column is poured and used without further equilibration.

   d. Development of the Chromatogram

   A 0.1-0.2 ml sample (10-20 mg) of freshly prepared hemolysate is dialyzed overnight against Developer BT-4 and then diluted to 1 ml with water before application. Development is made with

Developer BT-5 at 12 ml hr^{-1} with collection of 2.5 ml fractions.
After Hb γ_4 has eluted (usually less than 40 ml), the remaining
hemoglobin (Hb F, Hb A, or variants) is removed with Developer BT-6
as a single large fraction. Spectrophotometry is done at 415 nm.

2. *Examples*

The method was applied by Henson and Huisman (1978) to 63 samples
from Black newborns, some of whom were heterozygous for Hb S or Hb
C. Prior to chromatography, all samples had been electrophoresed
on starch gel at pH 9 and categorized into three groups on the basis
of apparent content of Hb γ_4: negative, trace, and positive. The
percentage in these three groups as determined by this method are
graphed in Figure 9.7.

3. *Comments*

The method provides a relatively rapid means for quantitative determinations of Hb γ_4.
        The results from negative samples (Figure 9.7) probably repre-
sent the absorbance from heme-containing nonhemoglobin proteins
such as catalase and are a background value for the other determinations.

B. Microchromatographic Method

The procedure for Hb F in the presence of Hb S and/or Hb C (Section
III.B above) is used with only minor variation. In this application the presence of Hb A does not interfere.

1. *Procedure*

Developer, sample, and column are exactly as in Section III.B,
above. Because the zone of Hb γ_4 is usually not visible, the first
4-5 ml of effluent are collected with care so that the collection
is stopped before any Hb F begins to emerge. The remaining hemoglobin is then removed into a separate flask with 2% KCN. Before
spectrophotometric determination at 415 nm, the first collection
which contains the Hb γ_4 is centrifuged to remove traces of stroma
and ion exchanger.

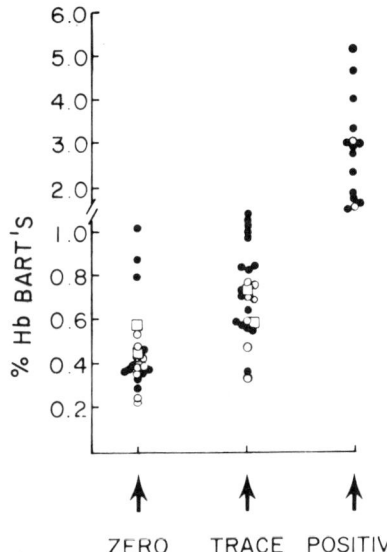

Figure 9.7. The quantities of hemoglobin Bart's (γ_4) in 63 Black newborn, determined by CM-cellulose chromatography. Open circles = babies with an Hb S heterozygosity; closed circles = babies with Hb A; open squares = babies with an Hb C heterozygosity. Division into three groups is based on the evaluation of the amount of Hb Bart's by inspection of a starch gel after separation of the hemoglobins at pH 9.0. (From Henson and Huisman, 1978, with permission.)

2. *Examples and Comments*

That the procedure, indeed, measures Hb γ_4 was assessed by using macrochromatography, by electrophoretic examination of chromatographic zones, and by amino acid analysis.

The procedure was applied to 136 unselected cord bloods which were also examined by starch-gel electrophoresis. In 118 samples in which Hb γ_4 was electrophoretically undetectable, the zone in the position of Hb γ_4 by the microchromatographic procedure ranged between 0.3 and 2.0% with an average of 0.9%. Eleven others had quantities between 2.2 and 3.5% and may represent the tail of a

skewed distribution curve.   However, in seven other samples, the
Hb γ_4 was 5.8, 6.1, 6.4, 7.0, 7.2, 8.4, and 9.9%.

The results by this very rapid micro method are comparable to
those by the macro method of the preceding section.

I. INTRODUCTION

In preceding chapters, the procedures that have been detailed are
designed mainly for the qualitative and quantitative determination
of the components in a mixture of hemoglobins. The sample in the
macrochromatographic methods on 1 X 15-60 cm columns is usually
30-50 mg. Consequently, even if the component is a minor one, say
an Hb A_2 variant in the amount of 1-2%, sufficient material can be
isolated to permit an amino acid analysis or an electrophoretic

study. Obviously, the higher the percentage the more component is
available for study by other methods. However, for numerous investigations much more of a particular hemoglobin may be desired, and,
consequently, preparative scale chromatograms have often been used.
Nevertheless, rarely is it necessary to use the hemoglobin from
500 ml of blood on a 10 X 100 cm column (Schroeder and Holmquist,
1966). Usually, if the column is 2-4 cm in diameter and 25-60 cm
long, a sufficient quantity for most subsequent studies can be isolated. Because of the many different reasons for which one might
want to isolate a hemoglobin in larger quantity, only some general
principles and examples can be presented.

Any conditions that have been used for the analytical chromatography of a hemoglobin form the basis for a preparative isolation.
Analytical chromatography should have determined not only the most
satisfactory type of ion exchanger for a separation, but also the
conditions of development. Indeed, the requirements of preparative
isolation are sufficiently different from those of analysis that the
conditions of development may be rather different. Thus, in preparative procedures, one need not elute the components successively
into the filtrate until the desired hemoglobin emerges; in fact, it
is most desirable not to do so. Once the analytical chromatogram
has given an idea of the behavior of the desired component, further
analytical chromatograms may be used to simplify conditions (such
as, for instance, from a gradient to a nongradient system), but
still adequately achieve the desired separations. Such a simpli-
fication, if possible, is most desirable mainly because the size of
gradients for larger columns is burdensome.

In analytical chromatography, it is unimportant that the hemoglobins are greatly diluted in the effluent; in preparative procedures, such dilution is a disadvantage because extensive concentration is then necessary. Consequently, it is advisable to arrange
the chromatographic conditions so that the desired hemoglobin is
separated from other hemoglobins but remains on the column. For
example, if the desired hemoglobin has a movement intermediate among

all the other components, the best strategy would be to select conditions so that the more rapidly moving hemoglobins are washed from the column and the desired hemoglobin is brought close to the bottom of the column. The chromatogram is then stopped while the upper portion of the column is removed. By passing a stronger developer (different pH or salt concentration) through the remaining portion of the column, the desired hemoglobin can be removed in a relatively concentrated solution. In another example, if the desired hemoglobin is the most strongly fixed, conditions would be chosen to move all other hemoglobins away from it. Then, after the chromatogram has been stopped, the portion with the desired component is removed, repoured in another tube, and removed with strong developer.

How much scaling-up of a chromatogram is needed in order to achieve the desired isolation? If an analytical column 1 cm in diameter can handle 50 mg hemoglobin, a column 3 cm in diameter and of equal length has 9 times the capacity and should be able to take 500 mg hemoglobin. Because it is unlikely that the full capacity of an analytical column is usually utilized, it may be possible in preparative work to chromatograph considerably more hemoglobin than one would calculate from the ratio of the sizes. If a gradient is necessary for the development of the preparative chromatogram, its volume compared to that for an analytical-scale column should at least approximate the ratio of the sizes of the columns. Of course, if a gradient is unnecessary, development is simply done with a single solution or a series of solutions until the appropriate separation is achieved.

It may not be worthwhile to increase the size of the column inordinately in order to chromatograph a very large sample. Rather, it may be better to divide the sample and use several smaller columns either concurrently or consecutively. Thus, the same results from the chromatography of the hemoglobin of 500 ml blood on a 10 X 100 cm column (Schroeder and Holmquist, 1966) could have been accomplished with twenty 3 X 35 cm columns. In retrospect, the one big column by no means gave the anticipated saving in time and effort.

On the following pages, examples will be given of the use of various ion exchangers and developmental conditions for the isolation of hemoglobins in larger quantities. Because the chromatographic behavior of the many known hemoglobin variants and the mixtures in which they occur are so varied, the comments above and the examples below can serve only as a guide to the selection of proper conditions for the preparative isolation of any component. If the chromatographic behavior of a hemoglobin is known, it is usually a simple matter to scale up the method for preparative work. The Appendix provides references to descriptions of preparative isolations of specific hemoglobin variants.

II. AMBERLITE IRC-50

Preparative isolation on Amberlite IRC-50 requires little more than a change in scale of the operations. The description below of the isolation of Hb F from cord blood is taken from Schroeder *et al.* (1963) and applies some of the principles noted above.

The preparation and equilibration of Amberlite IRC-50 as well as the composition of solutions is detailed in Chapter 5, Section II. After a 3.5 X 35 cm column of Amberlite IRC-50 had been equilibrated with 15 liters Developer 4 at 6°C, a 2-g sample of dialyzed cord-blood hemoglobin which had a concentration of about 60 mg ml^{-1} was applied to the column. Development with Developer 4 at a flow rate of 60 ml hr^{-1} was continued until minor component Hb F_I had emerged from the column and until Hb F was about to emerge. At this point the chromatogram was stopped, and the upper portion of the column with Hb A was removed. The remaining portion of the column with the Hb F was then warmed to 40°C for 10 min by circulating water through an outer jacket. Finally, the Hb F was removed with Developer 4 at 40°C in 10-15 min at a flow rate of 750 ml hr^{-1}. The effluent was immediately cooled in ice. In these experiments, the effluent with Hb F was concentrated by high speed centrifugation (Chapter 2, Section IX), but other means also described in Chapter 2 could have been used.

The above example took advantage of the rather high sensitivity to temperature that is exhibited in the chromatography of hemoglobins on Amberlite IRC-50. In the following instance, chromatography was done at room temperature with a different developer. The isolation of Hb A_{Ic} by Bunn *et al.* (1975) was based on the procedure of Trivelli, Ranney, and Lai (1971) for the quantitation of Hb A_{Ic} (Chapter 8, Section II). Trivelli, Ranney, and Lai (1971) used Amberlite IRC-50 at room temperature with Developer 6 instead of 6°C and Developer 4 as Allen, Schroeder, and Balog (1958) had done. Thus, Bunn *et al.* (1975) chromatographed as much as 5.5 g adult hemoglobin on a 5 X 50 cm column at room temperature with Developer 6. The Hb A_{Ic} was preceded through the column by Hb A_{Ia} and Hb A_{Ib} and was collected in the effluent. The authors do not describe the means for concentrating this probably very dilute solution of Hb A_{Ic}.

III. CARBOXYMETHYL-(CM)-CELLULOSE

Again, this preparative procedure is a scaled-up version of the quantitative analytical methods that were described in Chapter 5, where the data are for preparation of developers that are 0.01 M sodium phosphate buffers (pH 6.7-7.8) and for equilibration of the cation exchanger which is microgranular, preswollen CM-cellulose (CM-52). The size of the column and the volume of the developers to be used to form the pH gradient will be determined by the amount of hemoglobin to be isolated. It is the experience of the authors that about 200 mg hemoglobin can be successfully chromatographed on a 1.8 X 35 cm CM-cellulose column. Such a chromatogram is developed at room temperature with a flow rate as high as 40 ml hr^{-1}, and a pH gradient from a 500-ml constant-volume mixer that contains buffer at pH 6.7 and into which is fed buffer at pH 7.4. If necessary, the pH 7.4 developer is replaced after 24 hr by a developer with pH 7.6 or 7.8; the selection and sequence of these developers depends entirely on the type of hemoglobin variant to be isolated.

CM-cellulose preparative chromatography is useful for the isolation of larger quantities of the following variants.

A.   Hb H or β_4

This electrophoretically fast-moving variant is eluted in the first
40 ml under the conditions described above and is often contaminated
with nonhemoglobin proteins (NHP). After the appropriate zone has
been concentrated by vacuum dialysis, Hb H may be separated from
the NHP fraction by chromatography on a 2 X 60 cm column of Sephadex
G 75 in 0.01 M phosphate buffer, pH 7.4. The relative instability
of Hb H requires that all manipulations be done at 4°C.

B.   Hb Bart's or γ_4

This electrophoretically fast-moving variant elutes more slowly and
will separate more readily from Hb H and the NHP zone particularly
when the above conditions and a slow flow rate (not more than 20
ml hr^{-1}) are used. The solution containing this zone can be concentrated by vacuum dialysis. Although Hb Bart's is much more stable than Hb H, it is advisable to develop the chromatogram at 4°C.

C.   Slowly-moving Hemoglobin Variants

Although these hemoglobin variants are more easily isolated by
anion-exchange chromatography (see below), some of them can best
be isolated on columns of CM-cellulose because this procedure allows
the separation of Hb A_2 from those variants with electrophoretic
mobilities (at pH 8.6-9.0) similar to that of Hb A_2. These hemo-
globins are Hb C ($\alpha_2\beta_2$ 6 Glu → Lys), Hb C Siriraj ($\alpha_2\beta_2$ 7 Glu → Lys),
Hb British Columbia ($\alpha_2\beta_2$ 101 Glu → Lys), Hb O-Arab ($\alpha_2\beta_2$ 121 Glu →
Lys), Hb C-Harlem ($\alpha_2\beta_2$ 6 Glu → Val; 73 Asp → Asn), Hb Chad (α_2 23
Glu → Lys β_2), Hb O-Padova (α_2 30 Glu → Lys β_2), Hb O-Indonesia
(α_2 116 Glu → Lys β_2); however, Hb E-Saskatoon ($\alpha_2\beta_2$ 22 Glu → Lys)
and Hb E ($\alpha_2\beta_2$ 26 Glu → Lys) do not separate from Hb A_2 in this
system (Huisman and Wrightstone, 1974). Some of these variants
which have (nearly) identical electrophoretic mobilities can be
separated from each other by CM-cellulose chromatography; thus, Hb
C can be separated from Hb C-Harlem, Hb Agenogi, Hb O-Arab, or Hb
O-Indonesia. The chromatographic system is essentially a scaled-up

version of that which is detailed for analytical chromatography in
Chapter 5, Section IV.

IV. CARBOXYMETHYL-(CM)-SEPHADEX

The use of CM-Sephadex was once proposed to replace the CM-cellulose
method because of superior separations (see Chapter 5). However,
with the introduction of a more uniform type of CM-cellulose (the
microgranular CM-52, particularly) the advantage of using CM-Sephadex
has been eliminated except for the isolation of the several minor
hemoglobins. Thus, the following procedure can be used as an alternate method to the Amberlite IRC-50 procedure for the isolation of
the hemoglobins A_{Ia}, A_{Ib}, A_{Ic}, and A_{Id} (Dozy and Huisman, 1969).
The preparation of Tris-maleic acid buffers of various pH's
and the equilibration of CM-Sephadex have been described in Chapter
5, Section III.

About 500 mg hemoglobin from a freshly prepared hemolysate is
chromatographed on a 3.0 X 60 cm column of CM-Sephadex. The CM-
Sephadex is equilibrated with pH 6.5 buffer at room temperature but
chromatography is done at 4°C. To produce the pH gradient, pH 7.0
developer is introduced into a 2-liter constant-volume mixer which
has pH 6.5 buffer. The flow rate is 40 ml hr^{-1}. After approximately 3 liters of developer have passed through the column, the
remaining hemoglobin can be eluted by replacing the influent pH 7.0
buffer with pH 7.6 buffer.

Four to five minor components which are observed in a normal
freshly prepared hemolysate were eluted at pH's of 6.80, 6.85, 6.92,
and 7.01, respectively, as depicted in Figure 10.1. These components
have been identified as follows. I is a complex of Hb A and pyridoxal phosphate (Srivastava, van Loon, and Beutler, 1972). II and
III are Hb A_{Ia+b} and Hb A_{Ic} as shown in Figure 10.1d by the chromatographic behavior of Hb A_{Ia+b} and Hb A_{Ic} that had been isolated
by Amberlite IRC-50 chromatography (the contaminating Hb A in the
Hb A_{Ic} of part (d) probably is due to loss of the labile Schiff
base N-terminal blocking group of Hb A_{Ic}). IIIa contains mainly

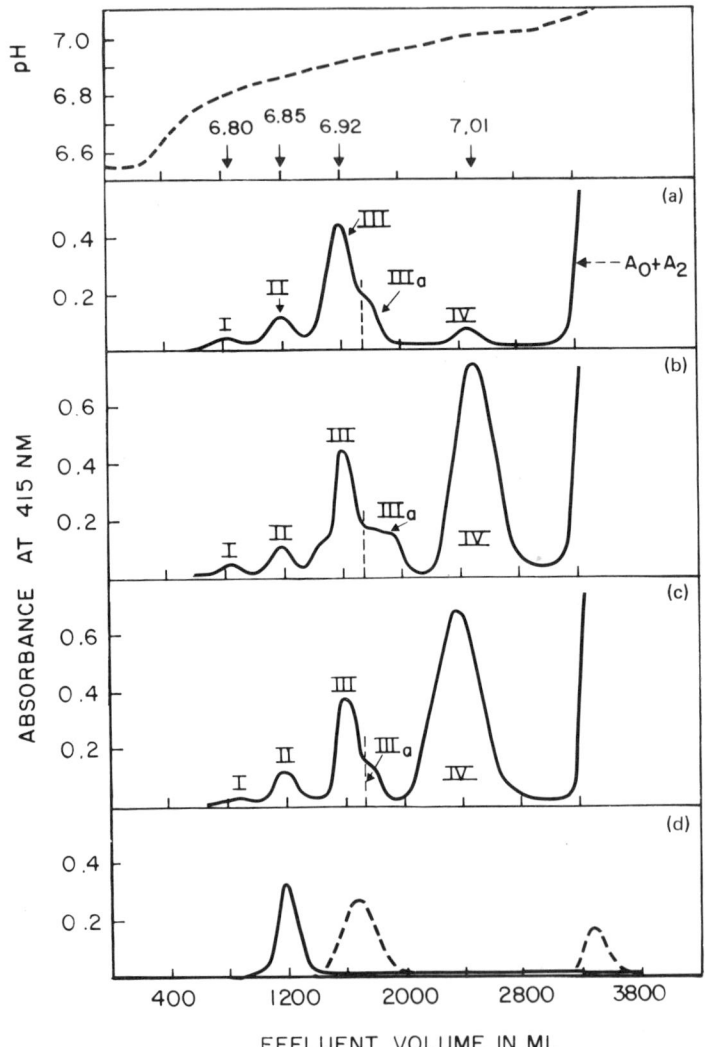

Figure 10.1. Separation of the minor hemoglobins by preparative CM-Sephadex chromatography. For further details, see text.

Hb F. Aging of red-cell hemolysate and/or incubation with oxidized glutathione (GSSG) results in a marked increase in component IV which, thus, may be a complex between Hb A and glutathione.

This procedure does not achieve the partial separation of Hb A_{Ia} and Hb A_{Ib} that occurs on Amberlite IRC-50. It does, however, incompletely separate Hb A_{Ic} and Hb F.

V. DIETHYLAMINOETHYL-(DEAE)-CELLULOSE

Just as analytical determinations on DEAE-cellulose may use several types of buffer for development so also may preparative procedures as described in the following sections.

A. Tris-HCl Developers

Chapter 6, Section III provides the necessary data for the preparation of the 0.05 M Tris-HCl developers and for the equilibration of the DEAE-cellulose (DE-52) which in this application should finally have equal volumes of settled resin and supernatant solution.

A 2.5 X 60 cm column is poured and equilibrated with pH 8.5 Tris-HCl solution for 2-3 hr at the rate of 50 ml hr^{-1}. About 2.5 g hemoglobin, which has been dialyzed overnight against the same developer, is applied to the column, and the chromatogram is developed with a pH 8.0 Tris-HCl developer at a flow rate of about 50 ml hr^{-1} at room temperature. Often the desired hemoglobin zone is not eluted, but a section with the hemoglobin of interest is removed, repoured into a separate glass tube, and is eluted with pH 7.0 Tris-HCl developer.

If it is desired to apply a pH gradient, the developer sequence is the same as described in Chapter 6 for the development of an analytical chromatogram, except that the volume of the mixer is increased to 500 ml. Because the volume of mixer is increased only two-fold and that of the chromatographic column by nine-fold, the gradient on the preparative chromatogram will be much steeper. The hemoglobins will emerge more rapidly and may be less completely separated.

Either procedure may be used for the isolation of both "fast-moving" and "slow-moving" abnormal hemoglobins and for Hb F which separates rather well from Hb A and Hb A_1 by this procedure.

B. Glycine-KCN Developers

As much as 500-1000 g DEAE-cellulose (DE-52) is equilibrated with
0.2 M glycine/0.01% KCN developer as described in Chapter 6, Section III and adjusted in pH to 7.85 with 4 M HCl with vigorous
stirring. Several 3 X 30 cm columns are poured and used simultaneously. Each column is loaded with 200-400 mg hemoglobin (in 2-5
ml) which preferably has been dialyzed overnight not against the
equilibrating developer but against 0.1 M glycine/0.01% KCN at 4°C.
If an undialyzed hemolysate is to be used, the sample is diluted
with twice its volume of 0.1 M glycine/0.01% KCN solution before
application to the column.

The chromatogram is developed at 50 ml hr^{-1} with an NaCl gradient at room temperature. The 1000-ml constant-volume mixer which
serves two columns simultaneously contains the 0.2 M glycine/0.01%
KCN/0.005 M NaCl developer, but the composition of the influent to
the mixer will depend on the hemoglobin variant to be isolated.
For instance, the NaCl concentration of this influent is 0.005 or
0.01 M for the isolation of Hb C ($\alpha_2\beta_2$ 6 Glu → Lys), Hb O-Arab
($\alpha_2\beta_2$ 121 Glu → Lys), Hb A_2; 0.01 M for the isolation of Hb S ($\alpha_2\beta_2$
6 Glu → Val) or Hb D ($\alpha_2\beta_2$ 121 Glu → Gln); 0.02 M for the isolation
of Hb J ($\alpha_2\beta_2$ 16 Gly → Asp); and 0.03 M for the isolation of Hb N
($\alpha_2\beta_2$ 95 Lys → Glu). Hemoglobin Hope ($\alpha_2\beta_2$ 136 Gly → Asp) which is
difficult to separate from Hb A can be isolated in pure form by a
sequence of developers in which the initial gradient between 0.005 M
and 0.01 M NaCl is changed after 24 hr (about 1000 ml); the 0.01 M
NaCl developer is replaced by the 0.02 M NaCl developer. Usually,
the desired hemoglobin is eluted from the appropriate section of
the ion exchanger after removal from the glass tube.

This procedure has recently been used with considerably success.
The isolated hemoglobins are purer than when isolated by many other
chromatographic methods.

The inconvenience of a gradient in preparative chromatograms
can often be avoided simply by scaling-up the appropriate microchromatographic procedures of Chapter 7. Thus, the following is an

example of the isolation of an unidentified α-chain variant (with
an alkaline electrophoretic behavior midway between that of Hb S and
Hb C) by means of a scaled-up version of the microchromatographic
procedure for Hb A_2 (Chapter 7, Section V.B). Thus, a 2.5 X 22 cm
column was packed with DEAE-cellulose (DE-52) that had been adjusted
in the supernatant solution to pH 7.12 in 0.2 M glycine/0.01% KCN
developer (see p. 111). The sample was the undialyzed hemolysate
of 3 ml washed, packed cells with 15 ml water and contained about
650 mg hemoglobin. Development at 100 ml hr^{-1} with 0.2 M glycine/
0.01% KCN for a total of 350 ml eluted Hb A_2 and the $\alpha_2^x\delta_2$ from the
column and moved the variant near the bottom of the column, separated from Hb A. After removal of the section with Hb A, the variant was eluted with 0.2 M glycine/0.2 M NaCl/0.01% KCN. For other
variants, the addition of some NaCl to the developer might be necessary; the appropriate conditions could easily be defined with
test chromatograms in straws (Chapter 7, Section VI).

C. Glycine Developers Without KCN

If it is desired to isolate an abnormal hemoglobin as oxyhemoglobin
for functional analyses, the following modification may be used.
The entire procedure has to be done at 4°C. The DE-52 is equilibrated with a large excess of a 0.2 M glycine solution (without KCN)
over a period of 48 hr during which the glycine solution is replaced
at least three times. The anion exchanger is used without adjustment of the pH. Onto a 3 X 30 cm column is applied 200-400 mg hemoglobin in 2-5 ml that has been dialyzed for 4-6 hr against 0.2 M
glycine at 4°C. The chromatogram is developed at 30 ml hr^{-1} for
1-2 hr first with an 0.2 M glycine/0.005 M NaCl solution. Then,
the developer is changed to 0.2 M glycine/0.01 M NaCl for the isolation of Hb C, Hb O, Hb D, Hb S, Hb A_2, Hb Lepore, etc., or to
0.2 M glycine/0.02 M NaCl for the isolation of Hb F and Hb Hope,
or to 0.2 M glycine/0.03 M NaCl for the isolation of Hb N-Baltimore.
Invariably, the section of the chromatogram that contains the desired hemoglobin is removed from the glass tube, poured into a

separate column, and eluted with an 0.2 M glycine/0.2 M NaCl solution. If the solution is concentrated by ultrafiltration, and dialyzed against an appropriate buffer for functional studies, the
isolated hemoglobin usually contains less than 5% ferrihemoglobin.

VI.   DIETHYLAMINOETHYL-(DEAE)-SEPHADEX

Preparative chromatography on DEAE-Sephadex has probably had more
widespread use than any other chromatographic procedure for hemoglobin isolation. It has been used for many years with much success
in the authors' laboratories. In principle, it can be said that,
if the pH of emergence of a hemoglobin on DEAE-Sephadex columns is
known from analytical chromatograms, that hemoglobin can best be
isolated from larger columns without a gradient if the ion exchanger
and eluting buffer are adjusted to a pH that is 0.3 units above the
pH of emergence. If many isolations are being made, it may not be
convenient to have DEAE-Sephadex that has been equilibrated to several different pH's. Consequently, equilibration can be made to
pH 8.5 and development with an appropriate buffer of lower pH. Details of the ion exchanger and the preparation of the developers
have been described in Chapter 6, Section II.

A 500-g portion of DEAE-Sephadex is mixed with 50 liters 0.05
M Tris-HCl, pH 8.5 (with 100 mg KCN liter^{-1}) and left for at least
24 hr. After the pH of the suspension of the swollen DEAE-Sephadex
has been adjusted with 1 M Tris solution to 8.3 to 8.4, the supernatant is replaced at least 5 times with pH 8.5 buffer.

Columns 3 X 60 cm in dimension are most suitable, although
larger columns can be used. Although undialyzed samples may be
used, dialysis against pH 8.5 buffer is advisable for the sample
which should contain not more than 1 g in 5-10 ml. The chromatogram is developed at about 50 ml hr^{-1} with a developer whose pH
depends upon the hemoglobin to be isolated. Thus, when electrophoretically slowly-moving variants such as hemoglobins C, O, D, S,
Lepore, and A$_2$ are to be isolated, the pH should be 8.0, but when
the variants are faster moving (hemoglobins J and N, and also Hb F),

the pH should be 7.85 to 7.90. When the desired separation has
been achieved, the section of the chromatogram with the desired
hemoglobin is removed, poured into another glass tube on top of a
small (5 cm) column of DEAE-Sephadex freshly equilibrated with 0.05
M Tris-HCl, pH 8.5, to prevent the slow elution of the hemoglobin,
and eluted with pH 6.8 developer. The isolated hemoglobin may be
concentrated by one of the several ways that are described in Chapter 2.

A combination of DEAE-Sephadex and DEAE-cellulose chromatography is ideally suited for the isolation of Hb F from the blood of
subjects whose hemoglobin has only small quantities of this fetal
protein. Although, as the analytical chromatogram in Figure 10.2
shows, DEAE-cellulose chromatography of the hemoglobin from a patient with a hereditary elevation of Hb F does separate the Hb F,
larger scale chromatography to isolate much Hb F in this way would
be difficult. Consequently, the Hb F from 100 ml of this patient's
blood was isolated in the following way. The available 14 g hemoglobin contained 2.2% Hb F (Figure 10.2) or about 310 mg. The entire 14 g was divided onto twenty 3 X 60 cm DEAE-Sephadex columns

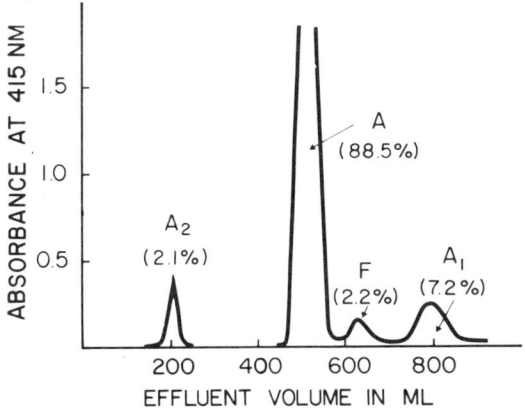

Figure 10.2. Analytical separation of the hemoglobin of a patient
with a slightly elevated level of Hb F by DEAE-cellulose chromatography according to the procedure of Chapter 6, Section III.

with pH 7.85 buffer as developer. The Hb A_1 + F zone (because
these two hemoglobins do not separate in this type of column chromatography) was removed, eluted, concentrated by ultrafiltration,
and dialyzed against 0.1 M glycine/0.1% KCN solution. The dialyzed
material was rechromatographed on six 3 X 28 cm columns of DEAE-
cellulose as described in Section V.B by applying a gradient between
0.005 and 0.02 M NaCl in 0.2 M glycine/0.01% KCN at a flow rate of
50 ml hr^{-1}. About 24 hr later, four distinct zones were observed
(Figure 10.3). The Hb F zone was removed from the glass tube, and

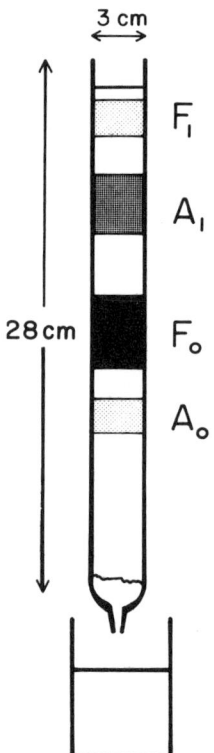

Figure 10.3. Separation of the hemoglobins A_o, F_o, A_1, and possibly
F_1 by DEAE-cellulose chromatography. The sample was isolated first
by preparative DEAE-Sephadex chromatography.

the Hb F was isolated from the anion exchanger with 0.2 M glycine/
0.01% KCN/0.2 M NaCl. The Hb F was 95% pure by starch-gel electrophoresis and amino acid analysis. The recovery was 220 mg or about
70%.

Chapter 11

CHROMATOGRAPHY OF ANIMAL HEMOGLOBINS

I. INTRODUCTION

Many studies of animal hemoglobins by both electrophoretic and chromatographic techniques have shown that hemoglobin of the majority of animal species is heterogeneous (for review and references, see Blunt, 1975; Kitchen, 1969; and Kitchen and Boyer, 1974). Thus, chromatographic procedures should be of great use for the quantitation and isolation of animal hemoglobins. Indeed, a rather superficial review of the literature verifies that several procedures which have been discussed in previous chapters are useful for these purposes. Table 11.1 provides an obviously incomplete survey of the application of chromatography to the study of hemoglobin in several animal species. It is not the intent to review these various applications in great detail; on the contrary only a few chromatographic experiments with some animal hemoglobins will be discussed to demonstrate the usefulness of these techniques.

Table 11.1 Some Animal Hemoglobins Studied by Different Chromatographic Procedures

Procedure	Species	References
Amberlite IRC-50	Primates	Buettner-Janusch, Buettner-Janusch, and Mason, 1970
	Cattle	Babin *et al.*, 1966; Schroeder *et al.*, 1967
	Chicken	Moss and Thompson, 1969; Saha, 1964
DEAE-Sephadex	Primates	Boyer *et al.*, 1971; Nute and Stamatoyannopoulas, 1971
	Chicken	Huisman and Schillhorn Van Veen, 1964; Huisman *et al.*, 1964
	Cattle	Dozy, Kleihauer, and Huisman, 1968; Schroeder *et al.*, 1972
	Sheep	Dozy, Kleihauer, and Huisman, 1968; Huisman *et al.*, 1968b; Huisman *et al.*, 1969
	Goat	Adams *et al.*, 1968; Dozy, Kleihauer, and Huisman, 1968; Huisman, 1970; Huisman *et al.*, 1969
	Deer	Huisman *et al.*, 1968a
	Shrimp	Waring, Poon, and Bowen, 1970
CM-Sephadex	Chicken	Brown and Ingram, 1974; Huisman and Schillhorn Van Veen, 1964; Huisman *et al.*, 1964
DEAE-cellulose	Rat	Garrick *et al.*, 1975
	Sheep	Blunt, 1975
CM-cellulose	Primates	Gandhi and Barnabas, 1970
	Chicken	Cirotto and Geraci, 1974; Godet, 1974; Hashimoto and Wilt, 1966
	Buffalo	Ranjekar and Barnabas, 1969a, 1969b
	Sheep	Huisman *et al.*, 1969
	Goat	Huisman, 1970
	Goose	Debouverie, 1975

II. APPLICATIONS

A. Amberlite IRC-50 Chromatography

This ion exchanger was used to separate adult hemoglobins A and B of cattle (Schroeder *et al.*, 1967). The general procedure was identical to that discussed in Chapter 5, Section II. However, the

developers for the separation of human hemoglobins were unsatisfactory for these two bovine hemoglobins. Developer 6 was replaced by
Developer 6c which contains 5.52 g $NaH_2PO_4 \cdot H_2O$, 0.71 g Na_2HPO_4, 0.65
g KCN liter^{-1} and has a pH of 6.50. For preparative purposes, 2 g
hemoglobin were chromatographed on a 3.5 X 35 cm column of Amberlite
IRC-50 at 6°C. The separation of the two hemoglobins is shown in

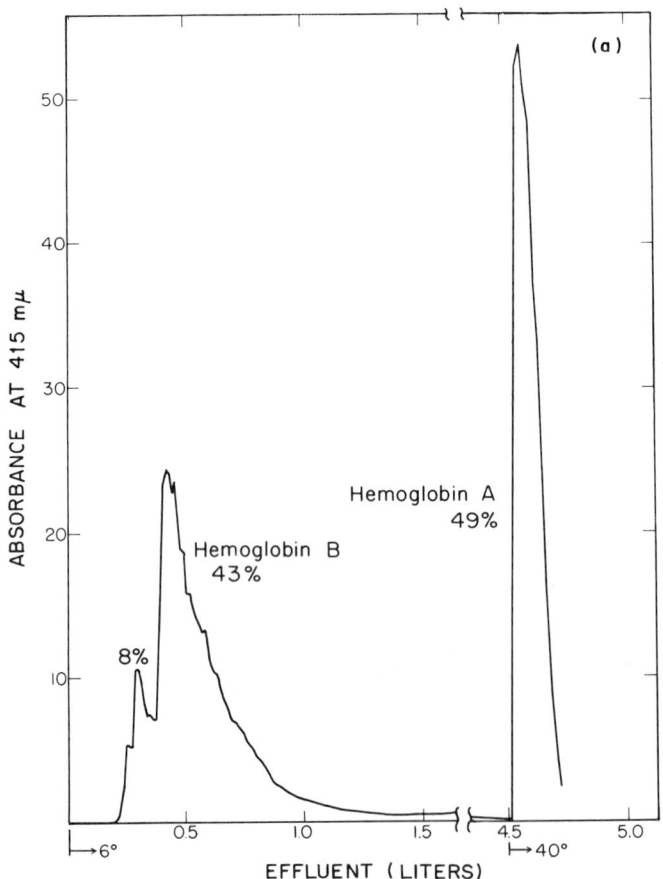

Figure 11.1. (a) The isolation of hemoglobins A and B from the
blood of a Jersey cow by chromatography on IRC-50. (From Schroeder
et al., 1967, with permission.) (b) The isolation of bovine hemoglobin F on Amberlite IRC-50 with Developer 6b. (From Babin *et al.*,
1966, with permission.)

Figure 11.1a. Hb B emerged rather rapidly but Hb A remained at the top of the column. This hemoglobin, however, was readily eluted when the column was warmed to 40°C. The quantities of hemoglobins A and B were about the same. The differences in chromatographic properties of the two proteins were undoubtedly due to the structural differences that exist between the β chains of Hb A and Hb B (positions 15-18-119: β^A chain Gly; Lys; Lys; β^B chain Ser; His; Asn).

A similar technique has been used to separate the fetal and adult hemoglobins from blood of a newborn calf (Babin *et al.*, 1966). A satisfactory developer was 6b which contains 6.9 g $NaH_2PO_4 \cdot H_2O$ and 0.65 g KCN liter^{-1}, and has a pH of 6.2. When about 1 g hemoglobin was chromatographed on a 3.5 X 30 cm column with Developer 6b at 6°C, three minor zones were eluted, the major zone (i.e., Hb F) was in the bottom half of the column, and the adult hemoglobin(s) was strongly fixed at the top of the column. Consequently,

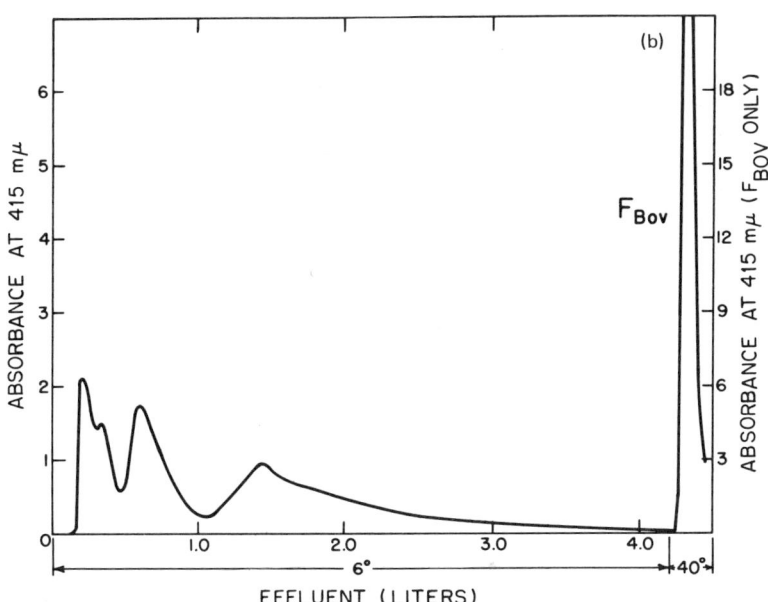

the section of the column with the adult hemoglobin was removed
from the glass tube, and the Hb F was eluted after the temperature
of the column had been raised to 40°C. Figure 11.1b shows the
chromatogram.

Thus, the Amberlite IRC-50 chromatographic method may readily
be adapted to separate hemoglobins with chromatographic properties
that are rather different from those of the human hemoglobins.
Other examples may be found in references quoted in Table 11.1.

B. Carboxymethyl-(CM)-cellulose
    Chromatography

The hemoglobin of the rat has great heterogeneity: six hemoglobins
have been separated by cellulose acetate electrophoresis. A study
of rat hemoglobin presents technical difficulties because one of
the several hemoglobins precipitates spontaneously *in vitro* at
physiological pH values. Examination of these various components,
however, has importance because some are synthesized at different
times during erythrocyte maturation.

Brada and Tobiska (1964) succeeded in separating these rat
hemoglobins by CM-cellulose chromatography. A 1.3 X 35 cm column
of CM-cellulose was equilibrated with an 0.01 M sodium phosphate
developer, pH 6.8, which contained 100 mg KCN liter^{-1}. In order to
prevent the crystallization of the hemoglobin, the freshly prepared
hemolysate was diluted to a concentration of 1 g dl^{-1}, dialyzed
against this developer for 20 min, and applied to the column as
soon as possible. About 80 mg was applied, and elution was achieved
with a pH gradient obtained by introducing a 0.01 M sodium phosphate
developer, pH 8.4 (with 100 mg KCN liter^{-1}) into the initial developer. The flow rate was maintained at 20 ml hr^{-1}. The separation
was excellent (Figure 11.2); at least six hemoglobins were observed.
When a 1% solution of the hemolysate was allowed to stand overnight,
centrifuged to remove crystals, and then chromatographed in the same
chromatographic system, zone A_3 was absent and presumably had crys-
tallized and was removed by centrifugation. The hemoglobin of the
newborn rat, although of a heterogeneity qualitatively similar to

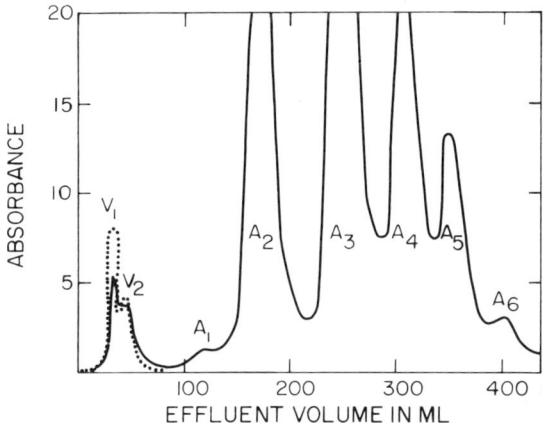

Figure 11.2. Chromatography of rat hemoglobin on a CM-cellulose column at room temperature. Dotted line indicates 280 nm absorbance; solid line indicates 410 nm absorbance. (From Brada and Tobiska, 1964, with permission.)

that of the adult rat, however, had a markedly different quantitative relationship between the various components.

More recently, Garrick *et al.* (1975) have observed a similar heterogeneity by means of DEAE-cellulose chromatography with 0.05 M Tris-HCl developers and a pH gradient.

Table 11.1 lists several other species whose hemoglobin and its heterogeneity have been studied by similar methods of CM-cellulose or CM-Sephadex chromatography.

C. Diethylaminoethyl-(DEAE)-cellulose
     and DEAE-Sephadex Chromatography

In addition to the two adult bovine hemoglobins A and B that were discussed earlier in this chapter, two other types, termed C and D, have been reported in a variety of breeds. All four hemoglobins have a different electrophoretic mobility and are readily separated by ion exchange chromatography. Figure 11.3a presents chromatograms on columns of DEAE-Sephadex by the method of Chapter 6, Section II. The structural differences which reside in the β chains (C → D at residue 20 Asp → Gly, at residue 43 Ser → Thr, and at residue 131

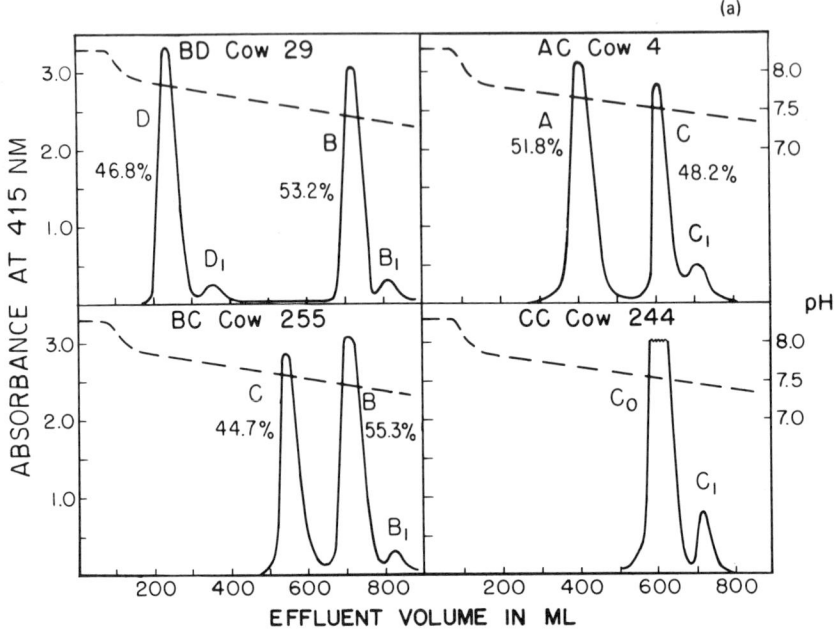

Figure 11.3. (a) Chromatographic separation of four bovine hemoglobins on columns of DEAE-Sephadex. The broken lines give the pH
values of the effluents. (From Schroeder *et al.*, 1972, with permission.) (b) Separation of the hemoglobin components of a newborn
goat which was heterozygous for the Hb $^{I\alpha B}$ allele on columns of
DEAE-Sephadex. (From Huisman *et al.*, 1969, with permission.)

Gln → Lys) explain satisfactorily the differences in the chromatographic properties of these two hemoglobins (Schroeder *et al.*, 1972).

Considerable heterogeneity has been observed in the hemoglobins
of sheep and goats. The various forms include three hemoglobins in
the adult sheep (termed A, B, and D) and five hemoglobins in the
adult goat (termed A, B, D, D-Malta, and E), which are found in
different combinations in individual animals of both species. Sheep
Hbs A and B and also goat Hbs A, D, D-Malta, and E differ in the β
chains, whereas sheep Hb D and goat Hb B differ from other types of
the same species in their α chains. Both species have a specific
fetal hemoglobin (Hb F) which disappears by about 40-50 days after
birth. The heterogeneity of the goat hemoglobin is further compli-

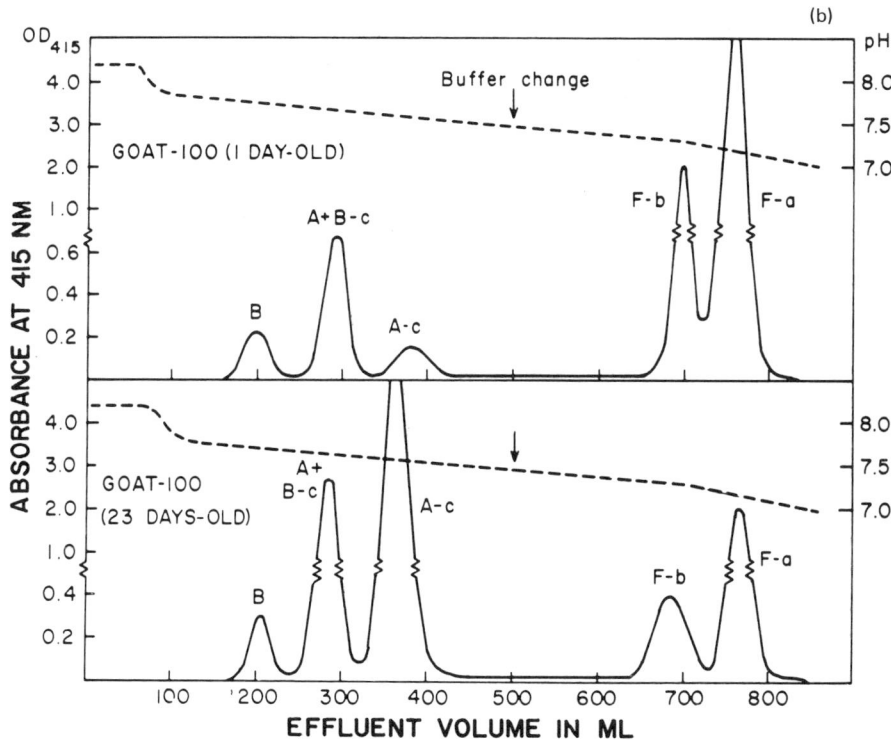

cated by the existence of two nonallelic α-chain structural genes, each of which produces a specific α chain ($^{I}α$ and $^{II}α$). Thus, each goat with Hb A has two forms of Hb A, namely $^{I}α_2β_2^{A}$ and $^{II}α_2β_2^{A}$. To complicate matters further, a dormant gene (the $β^{C}$ gene) in sheep and goats is activated during the first few months after birth as well as in animals who are subjected to severe anemia. This $β^{C}$ chain combines with the $^{I}α$ and $^{II}α$ chain to form Hb C. Table 11.2 is presented to help the reader comprehend the complexity of these hemoglobin heterogeneities.

Many of these hemoglobins, except those with $^{I}α$ or $^{II}α$ chains and identical β chains, can be separated by anion-exchange chromatography, mainly DEAE-Sephadex chromatography. As an example, the chromatograms of Figure 11.3b show the hemoglobins at the age of 1

Table 11.2 Major Hemoglobin Types Observed in Sheep and Goats

Species	Nonanemic animal	Severely anemic animal	Hb F in newborn
Sheep AA	A: $\alpha_2^A\beta_2^A$	C: $\alpha_2^A\beta_2^C$	$\alpha_2^A\gamma_2$
Sheep BB[a]	B: $\alpha_2^A\beta_2^B$	B: $\alpha_2^A\beta_2^B$	$\alpha_2^A\gamma_2$
Sheep AB	A+B	C+B	$\alpha_2^A\gamma_2$
Sheep AD	A: $\alpha_2^A\beta_2^A$ D: $\alpha_2^D\beta_2^A$	C: $\alpha_2^A\beta_2^C$ $\alpha_2^D\beta_2^C$	$\alpha_2^A\gamma_2$ $\alpha_2^D\gamma_2$
Goat AA	A: $^I\alpha_2\beta_2^A + {}^{II}\alpha_2\beta_2^A$	C: $^I\alpha_2\beta_2^C + {}^{II}\alpha_2\beta_2^C$	$^I\alpha_2\gamma_2 + {}^{II}\alpha_2\gamma_2$
Goat AD	A: $^I\alpha_2\beta_2^A + {}^{II}\alpha_2\beta_2^A$ D: $^I\alpha_2\beta_2^D + {}^{II}\alpha_2\beta_2^D$	C: $^I\alpha_2\beta_2^C + {}^{II}\alpha_2\beta_2^C$	$^I\alpha_2\gamma_2 + {}^{II}\alpha_2\gamma_2$
Goat BB	A: $^{II}\alpha_2\beta_2^A$	,C: $^{II}\alpha_2\beta_2^C$	$^{II}\alpha_2\gamma_2$
Goat AB	A: $^I\alpha_2\beta_2^A + {}^{II}\alpha_2\beta_2^A$ B: $^I\beta_2^B\beta_2^A$	C: $^I\alpha_2\beta_2^C + {}^{II}\alpha_2\beta_2^C$ $^I\beta_2^B\beta_2^C$	$^I\alpha_2\gamma_2 + {}^{II}\alpha_2\gamma_2$ $^I\beta_2^B\gamma_2$

[a]Sheep with Hb B appear not to have a dormant β^C gene, and thus Hb C is not found in anemic BB sheep.

and 23 days of a young kid who is heterozygous for the $^I\alpha$ chain variant Hb B (the $^I\alpha^B$ chain). Separation of the two fetal hemoglobins (the $^I\alpha_2^B\gamma_2$ component or F-b in Figure 11.3b and the $\alpha_2\gamma_2$ or F-a in Figure 11.3b which is a mixture of the $^I\alpha_2\gamma_2$ and $^{II}\alpha_2\gamma_2$ components) required the use of an 0.05 M Tris-HCl developer with a pH of 6.5. Hemoglobins containing the β^C chain were greatly increased in quantity when the kid was about 3 weeks old. The $\alpha_2\beta_2^C$ zone (A-c in Figure 11.3b) separated completely from the other zones; the $^I\alpha_2^B\beta_2^C$ component (B-c in Figure 11.3b), however, had exactly the same chromatographic properties in this system as Hb A. It is evident that anion-exchange chromatography is an important tool in the study of the hemoglobins of these animals. Although several of these hemoglobins can be completely separated from each other by this technique, on the basis of recent observations the DEAE-cellulose technique which uses glycine-NaCl-KCN developers (Chapter 6, Section III) will further improve the separations. The popularity of DEAE-Sephadex or DEAE-cellulose chromatography in the study of the animal hemoglobins is attested to by the size of the list in Table 11.1.

A good example of the use of an NaCl gradient in DEAE-Sephadex chromatography can be found in the study of the hemoglobin of the shrimp (Waring, Poon, and Bowen, 1970). The initial extract which contains the hemoglobins is greatly contaminated with acidic proteases which, however, can be separated from the hemoglobin by the DEAE-Sephadex chromatography (for details, see Waring, Poon, and Bowen, 1970). The separations by such means are depicted in Figure 11.4. The major hemoglobin (Hb-2) was eluted at a NaCl concentration of 0.19-0.20 M and Hb-1 at a slightly higher NaCl concentration (0.20-0.22 M). Electrophoretic examination of these components showed that a third hemoglobin (Hb-3) was present in the Hb-2 zone; Hb-3 is eluted slightly in front of the Hb-2, perhaps at 0.15-0.17 M NaCl. Denatured hemoglobin emerged before the three hemoglobins and residual material with brown and green colors remained at the top of the column. Acidic proteases and the green and brown material (likely hemoglobin that was denatured by proteolysis) were eluted at

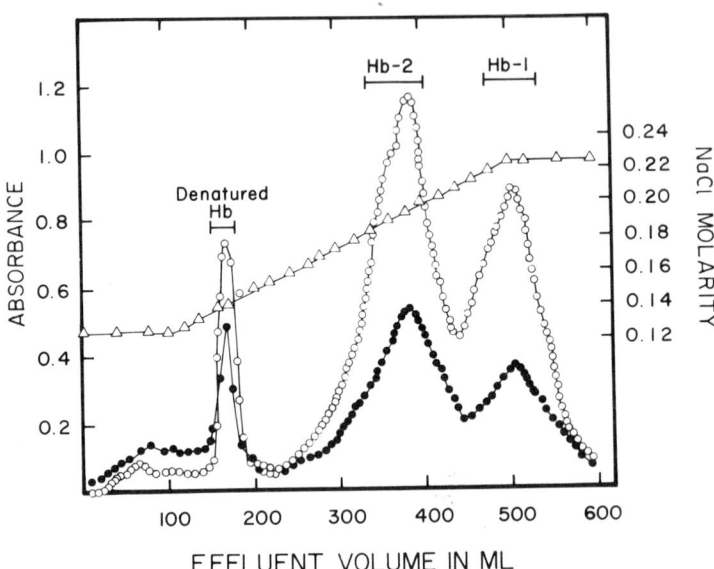

Figure 11.4. Chromatographic separation of Hb-1 and Hb-2 of the shrimp by DEAE-Sephadex chromatography. The sample (130 A_{415} units in 5 ml) was loaded onto a DEAE-Sephadex column (2.5 X 12 cm) and eluted first with 100 ml 0.125 M NaCl in Tris-HCl buffer at pH 7.5, then with a 400-ml linear NaCl gradient (0.125-0.225 M NaCl in the buffer), and finally with 100 ml 0.225 M NaCl (in the buffer). The flow rate was less than 5 ml hr^{-1} $(cm^2)^{-1}$. (From Waring, Poon, and Bowen, 1970, with permission.)

greater than 0.225 M NaCl. The chromatogram of Figure 11.4 is an excellent example of the possibilities that are offered by anion-exchange chromatographic procedures for the isolation of fish hemo-globin; perhaps, the replacement of DEAE-Sephadex by DEAE-cellulose and that of Tris-HCl-NaCl developers by glycine-NaCl-KCN developers will further improve the resolution.

Abraham, E. C., and Huisman, T. H. J. (1977). Differences in affinity of variant β chains for α chains. A possible explanation for the variation in the percentages of β chain variants in heterozygotes. *Hemoglobin 1*:861-873.

Abraham, E. C., Walker, D., Gravely, M., and Huisman, T. H. J. (1975). Minor hemoglobins in sickle cell anemia, β-thalassemia, and related conditions. A study of red cell fractions isolated by density gradient centrifugation. *Biochem. Med. 13*:56-77.

Abraham, E. C., Reese, A., Stallings, M., and Huisman, T. H. J. (1976-1977). Separation of human hemoglobins by DEAE-cellulose chromatography using glycine-KCN-NaCl developers. *Hemoglobin 1*:27-44.

Abraham, E. C., Huisman, T. H. J., Schroeder, W. A., Pace, L. A., and Grussing, L. (1977a). Microchromatography of hemoglobins. VII. Detection of some uncommon hemoglobin variants and two rapid methods for the quantitation of Hb-A_2 in the presence of Hb-C. *J. Chromatog. 143*:57-63.

Abraham, E. C., Reese, A., Stallings, M., Garver, F. A., and Huisman, T. H. J. (1977b). An improved chromatographic procedure for quantitation of human fetal hemoglobin. *Hemoglobin 1*:547-560.

Abraham, E. C., Huff, T. A., Cope, N. D., Wilson, J. B., Jr., Bransome, E. D., Jr., and Huisman, T. H. J. (1978). Determination of the glycosylated hemoglobins (Hb A_I) with a new microcolumn procedure. *Diabetes 27*:931-937.

Adams, H. R., Boyd, E. M., Wilson, J. B., Miller, A., and Huisman, T. H. J. (1968). The structure of goat hemoglobins. III. Hemoglobin D, a β chain variant with one apparent amino acid substitution (21 Asp → His). *Arch. Biochem. Biophys. 127*:398-405.

Allen, D. W., Schroeder, W. A., and Balog, J. (1958). Observations on the chromatographic heterogeneity of normal adult and fetal human hemoglobin. A study of the effects of crystallization and chromatography on the heterogeneity and isoleucine content. *J. Am. Chem. Soc. 80*:1628-1634.

Alm, R. S., Williams, R. J. P., and Tiselius, A. (1952). Gradient elution analysis. I. A general treatment. *Acta Chem. Scand. 6*:826-836.

Anonymous (1977). Glycosylated haemoglobins and disease. *Lancet*, July 2, pp. 22-23.

Antonini, E., Wyman, J., Brunori, M., Fronticelli, C., Bucci, E., and Rossi-Fanelli, A. (1965). Studies on the relations between molecular and functional properties of hemoglobin. V. The influence of temperature on the Bohr effect in human and in horse hemoglobin. *J. Biol. Chem. 240*:1096-1103.

Babin, D. R., Schroeder, W. A., Shelton, J. R., Shelton, J. B., and Robberson, B. (1966). The amino acid sequence of the γ chain of bovine fetal hemoglobin. *Biochemistry 5*:1297-1310.

Bernini, L. F. (1969). Rapid estimation of hemoglobin A_2 by DEAE chromatography. *Biochem. Genet. 2*:305-310.

Betke, K., Marti, H. R., and Schlicht, I. (1959). Estimation of small percentages of foetal haemoglobin. *Nature* (Lond.) *184*:1877-1878.

Blunt, M. H. (ed.) (1975). *The Blood of Sheep—Composition and Function*. Springer-Verlag, New York.

Boardman, N. K., and Partridge, S. M. (1953). Separation of neutral proteins on ion-exchange resins. *Nature* (Lond.) *171*:208-210.

Boardman, N. K., and Partridge, S. M. (1955). Separation of neutral proteins on ion-exchange resins. *Biochem. J. 59*:543-552.

Bock, R. M., and Ling, N.-S. (1954). Devices for gradient elution in chromatography. *Anal. Chem. 26*:1543-1546.

Bookchin, R. M., and Gallop, P. M. (1968). Structure of hemoglobin A_{Ic}. Nature of the N-terminal β chain blocking group. *Biochem. Biophys. Res. Commun. 32*:86-93.

Boyer, S. H., Crosby, E. F., Noyes, A. N., Fuller, G. F., Leslie, S. E., Donaldson, L. J., Vrablik, G. R., Schaefer, E. W., Jr., and Thurmon, T. F. (1971). Primate hemoglobins. Some sequences and some proposals concerning the character of evolution and mutation. *Biochem. Genet. 5*:405-448.

Boyer, S. H., Boyer, M. L., Noyes, A. N., and Balding, T. K. (1974). Immunological basis for detection of sickle cell hemoglobin phenotypes in amniotic fluid erythrocytes. *Ann. N.Y. Acad. Sci. 241*:699-713.

Brada, Z., and Tobiska, J. (1964). Host-tumour relationship. XVI. Heterogeneity of rat haemoglobin. *Neoplasma 11*:371-378.

Brennan, S. O., Winterbourn, C. C., and Carrell, R. W. (1977). Isolation of high oxygen affinity hemoglobins. *Hemoglobin 1*: 479-485.

Brown, J. L., and Ingram, V. M. (1974). Structural studies on chick embryonic hemoglobins. *J. Biol. Chem. 249*:3960-3972.

Bucci, E., and Fronticelli, C. (1965). A new method for the preparation of α and β subunits of human hemoglobin. *J. Biol. Chem. 240*:PC551-552.

Buettner-Janusch, V., Buettner-Janusch, J., and Mason, G. A. (1970). Multiple haemoglobins of mandrills, *Papio sphinx*. *Int. J. Biochem. 1*:322-326.

Bunn, H. F., Forget, B. G., and Ranney, H. M. (1977). *Human Hemoglobins*. Saunders, Philadelphia.

Bunn, H. F., Gabbay, K. H., and Gallop, P. M. (1978). The glycosylation of hemoglobin. Relevance to diabetes mellitus. *Science 200*:21-27.

Bunn, H. F., Haney, D. N., Gabbay, K. H., and Gallop, P. M. (1975). Further identification of the nature and linkage of the carbohydrate in hemoglobin A_{Ic}. *Biochem. Biophys. Res. Commun. 67*: 103-109.

Bunn, H. F., Haney, D. N., Kamin, S., Gabbay, K. H., and Gallop,
    P. M. (1976). The biosynthesis of human hemoglobin A_{Ic}. Slow
    glycosylation of hemoglobin *in vivo*. *J. Clin. Invest*. *57*:1652-
    1659.

Cerami, A., Koenig, R., and Peterson, C. M. (1978). Haemoglobin
    A_{Ic} and diabetes mellitus. *Br. J. Haematol*. *38*:1-4.

Chudwin, D. S., and Rucknagel, D. L. (1974). Immunological quanti-
    fication of hemoglobins F and A_2. *Clin. Chim. Acta 50*:413-418.

Cirotto, C., and Geraci, G. (1974). Exposed sulphydryl groups of
    chicken haemoglobins. Globin localization and effect of oxy-
    genation on their reactivity. *J. Molec. Biol*. *84*:103-114.

Clegg, M. D., and Schroeder, W. A. (1959). A chromatographic study
    of minor components of normal adult human hemoglobin including
    a comparison of hemoglobin from normal and phenylketonuric in-
    dividuals. *J. Am. Chem. Soc*. *81*:6065-6069.

Cole, R. A., Soeldner, J. S., Dunn, P. J., and Bunn, H. F. (1978).
    A rapid method for the determination of glycosylated hemoglo-
    bins using high pressure liquid chromatography. *Metabolism
    27*:289-301.

Davis, J. E., McDonald, J. M., and Jarett, L. (1977). A high per-
    formance liquid chromatography (HPLC) method for hemoglobin
    A_{Ic}. *Diabetes* (Suppl. 1) *26*:368.

Debouverie, D. (1975). Structure primaire de la chaine α du com-
    posant majeur de l'hemoglobine d'oie (Anser anser). *Biochimie
    57*:569-578.

Dozy, A. M., and Huisman, T. H. J. (1969). Studies on the heter-
    ogeneity of hemoglobin. XIV. Chromatography of normal and
    abnormal human hemoglobin types on CM-Sephadex. *J. Chromatog*.
    40:62-70.

Dozy, A. M., Kleihauer, E. F., and Huisman, T. H. J. (1968). Studies
    on the heterogeneity of hemoglobin. XIII. Chromatography of
    various human and animal hemoglobin types on DEAE-Sephadex.
    J. Chromatog. *32*:723-727.

Efremov, G. D., and Huisman, T. H. J. (1974). Studies on the heter-
    ogeneity of hemoglobin. XV. Separation of fetal hemoglobin

and the normally occurring minor adult hemoglobins by chromatography on DEAE-cellulose. *J. Chromatog. 89*:191-196.

Efremov, G. D., and Huisman, T. H. J. (1975). Further studies on
the use of microchromatography in mass-testing programs for
hemoglobinopathies. Proceedings of the Symposium on Sickle
Cell Anemia, Abidjan, 1975. *INSERM 44*:131-140.

Efremov, G. D., Huisman, T. H. J., Bowman, K., Wrightstone, R. N.,
and Schroeder, W. A. (1974a). Microchromatography of hemoglobins. II. A rapid method for the determination of hemoglobin
A_2. *J. Lab. Clin. Med. 84*:657-664.

Efremov, G. D., Wrightstone, R. N., Braun, R. C., Brodie, A. N.,
Mayson, S., and Huisman, T. H. J. (1974b). The use of micro-
chromatography in mass-testing programs for hemoglobinopathies.
Int. Istanbul Symp. Abnorm. Hemoglob. Thalass., Istanbul,
August, 1974, pp. 237-255.

Fahey, J. L., and Goodman, H. C. (1960). Characterization of anti-
thyroglobulin factors in human serum. *J. Clin. Invest. 39*:
1259-1265.

Fahey, J. L., McCoy, P. F., and Goulian, M. (1958). Chromatography
of serum proteins in normal and pathologic sera. The distribution of protein-bound carbohydrate and cholesterol, sidero-
philin, thyroxin-binding protein, B_{12}-binding proteins, alkaline and acid phosphatases, radioiodinated albumin and myeloma
proteins. *J. Clin. Invest. 37*:272-284.

Fitzgibbons, J. F., Koler, R. D., and Jones, R. T. (1976). Red
cell age-related changes of hemoglobins A_{Ia+b} and A_{Ic} in normal and diabetic subjects. *J. Clin. Invest. 58*:820-824.

Flückiger, R., and Winterhalter, K. H. (1976). *In vitro* synthesis
of hemoglobin A_{Ic}. *FEBS Lett. 71*:356-360.

Gale, R. E., Clegg, J. B., and Huehns, E. R. (1979). Human embryonic haemoglobins Gower 1 and Gower 2. *Nature* (Lond.) *280*:
162-164.

Gandhi, N. S., and Barnabas, J. (1970). Comparative studies of
human and langur hemoglobins. *Ind. J. Biochem. 1*:76-77.

Garel, M. C., Cohen-Solal, M., Blouquit, Y., and Rosa, J. (1974).

A method for isolation of abnormal haemoglobins with high oxygen affinity due to a frozen quaternary R-structure. Application to Hb Creteil $\alpha_2^A\beta_2$ (F5) 89 Asn. *FEBS Lett.* *43*:93-96.

Garrick, L. M., Sharma, V. S., McDonald, M. J., and Ranney, H. M. (1975). Rat haemoglobin heterogeneity. *Biochem. J.* *149*:245-258.

Garver, F. A., Jones, C. S., Baker, M. M., Altay, G., Barton, B. P., Gravely, M., and Huisman, T. H. J. (1976). Specific radioimmunochemical identification and quantitation of hemoglobins A_2 and F. *Am. J. Hematol.* *1*:459-469.

Godet, J. (1974). Hb F synthesis in chicken embryonic and postnatal development. Studies in various explanted erythropoietic tissues. *Devel. Biol.* *40*:199-207.

Hashimoto, K., and Wilt, F. H. (1966). The heterogeneity of chicken hemoglobin. *Proc. Natl. Acad. Sci. USA* *56*:1477-1483.

Hayashi, A., Suzuki, T., Shimizu, A., and Yamamura, Y. (1968). Properties of hemoglobin M. Unequivalent nature of the α and β subunits in the hemoglobin molecule. *Biochim. Biophys. Acta* *168*:262-273.

Henson, J., and Huisman, T. H. J. (1978). Possible relationship between the level of Hb Bart's (γ_4) and the relative amount of Hb S or Hb C in black heterozygous newborn. *Hemoglobin* *2*:393-398.

Hill, R. J., Konigsberg, W., Guidotti, G., and Craig, L. C. (1962). The structure of human hemoglobin. I. The separation of the α and β chains and their amino acid composition. *J. Biol. Chem.* *237*:1549-1554.

Hirs, C. H. W., Moore, S., and Stein, W. H. (1953). A chromatographic investigation of pancreatic ribonuclease. *J. Biol. Chem.* *200*:493-506.

Holmquist, W. R., and Schroeder, W. A. (1966). A new N-terminal blocking group involving a Schiff base in hemoglobin A_{Ic}. *Biochemistry* *5*:2489-2503.

Horton, B. F., and Chernoff, A. I. (1970). Miniature column chromatography of hemoglobins. *J. Chromatog.* *47*:493-498.

Horton, B. F., and Huisman, T. H. J. (1965). Studies on the heterogeneity of haemoglobin. VII. Minor haemoglobin components
in haematological diseases. *Br. J. Haematol.* 11:296-304.

Huisman, T. H. J. (1961). Quantitative determination of Hb A_2 using
DEAE-cellulose chromatography. *Proc. Hemoglob. Conf.*, Vienna,
pp. 95-96.

Huisman, T. H. J. (1970). Multiple α and β chain structural genes
as a basis for hemoglobin heterogeneity of the adult goat. In
Protides of the Biological Fluids, 17th Colloquium, Bruges,
1969 (H. Peeters, ed.). Pergamon, Oxford, pp. 241-248.

Huisman, T. H. J. (1972a). Normal and abnormal human hemoglobins.
Adv. Clin. Chem. 15:149-253.

Huisman, T. H. J. (1972b). Chromatographic separation of hemoglobins A_2 and C. The quantities of hemoglobin A_2 in patients
with AC trait, CC disease, and C-β-thalassemia. *Clin. Chim.
Acta* 40:159-163.

Huisman, T. H. J. (1977). Trimodality in the percentages of β chain
variants in heterozygotes. The effect of the number of active
Hb_α structural loci. *Hemoglobin* 1:349-382.

Huisman, T. H. J., and Dozy, A. M. (1961). Quantitative determination of the minor hemoglobin component Hb-A_2 by DEAE-cellulose
chromatography. *Anal. Biochem.* 2:400-403.

Huisman, T. H. J., and Dozy, A. M. (1962a). Studies on the heterogeneity of hemoglobin. IV. Chromatographic behavior of different human hemoglobins on anion-exchange cellulose (DEAE-
cellulose). *J. Chromatog.* 7:180-203.

Huisman, T. H. J., and Dozy, A. M. (1962b). Studies on the heterogeneity of hemoglobin. V. Binding of hemoglobin with oxidized
glutathione. *J. Lab. Clin. Med.* 60:302-319.

Huisman, T. H. J., and Dozy, A. M. (1965). Studies on the heterogeneity of hemoglobin. IX. The use of tris(hydroxymethyl)-
aminomethane-HCl buffers in the anion-exchange chromatography
of hemoglobins. *J. Chromatog.* 19:160-169.

Huisman, T. H. J., and Horton, B. F. (1965). Studies on the heterogeneity of hemoglobin. VIII. Chromatographic and electropho-

retic investigations of various minor hemoglobin fractions
present in normal and *in vitro* modified red blood cell hemolysates. *J. Chromatog. 18*:116-123.

Huisman, T. H. J., and Jonxis, J. H. P. (1977). *The Hemoglobinopathies—Techniques of Identification.* Dekker, New York.

Huisman, T. H. J., and Meyering, C. A. (1960). Studies on the
heterogeneity of hemoglobin. I. The heterogeneity of different human hemoglobin types in carboxymethylcellulose and
in Amberlite IRC-50 chromatography: Qualitative aspects.
Clin. Chim. Acta 5:103-123.

Huisman, T. H. J., and Prins, H. K. (1955). Chromatographic estimation of four different human hemoglobins. *J. Lab. Clin.
Med. 46*:255-262.

Huisman, T. H. J., and Prins, H. K. (1957). The chromatographic
behavior of different abnormal human haemoglobins on the
cation exchanger Amberlite IRC-50. *Clin. Chim. Acta 2*:307-311.

Huisman, T. H. J., and Schillhorn Van Veen, J. M. (1964). Studies
on animal hemoglobins. III. The possible role of intercellu-
lar inorganic phosphate on the oxygen equilibrium of the hemoglobin in the developing chicken. *Biochim. Biophys. Acta 88*:
367-374.

Huisman, T. H. J., and Wrightstone, R. N. (1974). Studies on the
heterogeneity of hemoglobin. XVI. Separation of variants
with a Glu → Lys substitution by chromatography on CM-cellulose.
J. Chromatog. 92:391-399.

Huisman, T. H. J., Martis, E. A., and Dozy, A. (1958). Chromatography of hemoglobin types on carboxymethylcellulose. *J. Lab.
Clin. Med. 52*:312-327.

Huisman, T. H. J., Schillhorn Van Veen, J. M., Dozy, A. M., and
Nechtman, C. M. (1964). Studies on animal hemoglobins. II.
The influence of inorganic phosphate on the physico-chemical
and physiological properties of the hemoglobin of the adult
chicken. *Biochim. Biophys. Acta 88*:352-366.

Huisman, T. H. J., Dozy, A. M., Horton, B. F., and Nechtman, C. M.
(1966). Studies on the heterogeneity of hemoglobin. X. The

nature of various minor hemoglobin components produced in human red blood cell hemolysates on aging. *J. Lab. Clin. Med. 67*: 355-373.

Huisman, T. H. J., Dozy, A. M., Blunt, M. H., and Hayes, F. A. (1968a). The hemoglobin heterogeneity of the Virginia white-tailed deer. A possible genetic explanation. *Arch. Biochem. Biophys. 127*:711-717.

Huisman, T. H. J., Dozy, A. M., Wilson, J. B., Efremov, G. D., and Vaskov, B. (1968b). Sheep hemoglobin D, an α-chain variant with one apparent amino acid substitution (α 15 Gly $\rightarrow$ Asp). *Biochim. Biophys. Acta 160*:467-469.

Huisman, T. H. J., Lewis, J. P., Blunt, M. H., Adams, H. R., Miller, A., Dozy, A. M., and Boyd, E. M. (1969). Hemoglobin C in newborn sheep and goats. A possible explanation for its function and biosynthesis. *Pediatr. Res. 3*:189-198.

Huisman, T. H. J., Schroeder, W. A., Brodie, A. N., Mayson, S. M., and Jakway, J. (1975). Microchromatography of hemoglobins. III. A simplified procedure for the determination of hemo-globin A_2. *J. Lab. Clin. Med. 86*:700-702.

Itano, H. A., and Robinson, E. A. (1960). Genetic control of the α- and β-chains of hemoglobin. *Proc. Natl. Acad. Sci. USA 46*:1492-1501.

Jones, R. T., and Schroeder, W. A. (1963a). Chromatography of human hemoglobin. Factors influencing chromatography and dif-ferentiation of similar hemoglobins. *J. Chromatog. 10*:421-431.

Jones, R. T., and Schroeder, W. A. (1963b). Chemical characteriza-tion and subunit hybridization of human hemoglobin H and associ-ated compounds. *Biochemistry 2*:1357-1367.

Jones, M. B., Koler, R. D., and Jones, R. T. (1978). Micro-column method for the determination of hemoglobin minor fractions A_{Ia+b} and A_{Ic}. *Hemoglobin 2*:53-58.

Kitchen, H. (1969). Heterogeneity of animal hemoglobins. *Adv. Vet. Sci. Comp. Med. 13*:247-330.

Kitchen, H., and Boyer, S. H. (eds.) (1974). Hemoglobin. Comparative molecular biology models for the study of disease. *Ann. N.Y. Acad. Sci. 241*:1-737.

Koenig, R. J., Blobstein, S. H., and Cerami, A. (1977). Structure of carbohydrate of hemoglobin A$_{\text{Ic}}$. *J. Biol. Chem. 252*:2992-2997.

Koenig, R. J., Peterson, C. M., Jones, R. L., Saudek, C., Lehrman, M., and Cerami, A. (1976). Correlation of glucose regulation and hemoglobin A$_{\text{Ic}}$ in diabetes mellitus. *N. Engl. J. Med. 295*:417-420.

Konigsberg, W., and Lehmann, H. (1965). The amino acid substitution in hemoglobin M$_{\text{Iwate}}$. *Biochim. Biophys. Acta 107*:266-269.

Krishnamoorthy, R., Gacon, G., and Labie, D. (1977). Isolation and partial characterization of hemoglobin A$_{\text{Ib}}$. *FEBS Lett. 77*: 99-102.

Kynoch, P. A. M., and Lehmann, H. (1977). Rapid estimation ($2\frac{1}{2}$ hours) of glycosylated haemoglobin for routine purposes. *Lancet*: July 2, p. 16.

Lehmann, H., and Huntsman, R. G. (1974). *Man's Haemoglobins*, 2nd ed. North-Holland, Amsterdam.

Lutcher, C. L., Wilson, J. B., Gravely, M. E., Stevens, P. D., Chen, C. J., Lindeman, J. G., Wong, S. C., Miller, A., Gottlieb, M., and Huisman, T. H. J. (1976). Hb Leslie, an unstable hemoglobin due to deletion of glutaminyl residue β 131 (H9) occurring in association with β^0-thalassemia, Hb C, and Hb S. *Blood 47*:99-112.

Matsuda, G., Schroeder, W. A., Jones, R. T., and Weliky, N. (1960). Is there an "embryonic" or "primitive" human hemoglobin? *Blood 16*:984-996.

McDonald, M. J., Shapiro, R., Bleichman, M., Solway, J., and Bunn, H. F. (1978). Glycosylated minor components of human adult hemoglobin. Purification, identification, and partial structural analysis. *J. Biol. Chem. 253*:2327-2332.

Meyering, C. A., Israels, A. L. M., Sebens, T., and Huisman, T. H. J. (1960). Studies on the heterogeneity of hemoglobin. II. The heterogeneity of different human hemoglobin types in carboxy-

methylcellulose and in Amberlite IRC-50 chromatography: Quantitative aspects. *Clin. Chim. Acta* 5:208-222.

Morrison, M., and Cook, J. L. (1955). Chromatographic fractionation
of normal adult oxyhemoglobin. *Science* 122:920-921.

Morrison, M., and Cook, J. L. (1957). Column chromatography of
human hemoglobins. *Fed. Proc.* 16:763-766.

Moss, B. A., and Thompson, E. O. P. (1969). Haemoglobins of the
adult domestic fowl, *Gallus domesticus*. *Aust. J. Biol. Sci.*
22:1455-1471.

Nute, P. E., and Stamatoyannopoulos, G. (1971). The ontogenesis of
hemoglobins in *Macaca nemestrina*. *Blood* 38:108-115.

Pauling, L., Itano, H. A., Singer, S. J., and Wells, I. C. (1949).
Sickle cell anemia, a molecular disease. *Science* 110:543-548.

Peterson, C. M., and Jones, R. L. (1977). Minor hemoglobins, diabetic "control," and diseases of postsynthetic protein modification. *Ann. Int. Med.* 87:489-491.

Peterson, E. A., and Sober, H. A. (1956). Chromatography of proteins. I. Cellulose ion-exchange adsorbents. *J. Am. Chem.
Soc.* 78:751-755.

Peterson, E. A., and Sober, H. A. (1959). Variable gradient device
for chromatography. *Anal. Chem.* 31:857-862.

Powars, D., Schroeder, W. A., and White, L. (1975). Rapid diagnosis
of sickle cell disease at birth by microcolumn chromatography.
Pediatrics 55:630-635.

Prins, H. K. (1959). The separation of different types of human
haemoglobin. *J. Chromatog.* 2:445-486.

Prins, H. K., and Huisman, T. H. J. (1955). Chromatographic estimation of different kinds of human haemoglobin. *Nature* (Lond.)
175:903-904.

Prins, H. K., and Huisman, T. H. J. (1956). Chromatographic beha-
viour of haemoglobin E. *Nature* (Lond.) 177:840-841.

Ranjekar, P. K., and Barnabas, J. (1969a). Haemoglobin phenotypes
in water buffalo (*Bos bubalus*) during development. *Comp.
Biochem. Physiol.* 28:1395-1401.

Ranjekar, P. K., and Barnabas, J. (1969b). Comparative aspects of developmental haemoglobins in ruminants. *Ind. J. Biochem. 6:* 1-5.

Ranney, H. M. (ed.) (1974). Hemoglobinopathies. *Sem. Haematol.* *11:*383-601.

Ranney, H. M., Nagel, R. L., Heller, P., and Udem, L. (1968). Oxygen equilibrium of hemoglobin M$_{Hyde Park}$. *Biochim. Biophys. Acta 160:*112-115.

Saha, A., (1964). Comparative studies on chick hemoglobins. *Biochim. Biophys. Acta 93:*573-584.

Schleider, C. T. H., Mayson, S. M., and Huisman, T. H. J. (1977). Further modification of the microchromatographic determination of hemoglobin A$_2$. *Hemoglobin 1:*503-504.

Schnek, A. G., and Schroeder, W. A. (1961). The relation between the minor components of whole normal human adult hemoglobin as isolated by chromatography and starch block electrophoresis. *J. Am. Chem. Soc. 83:*1472-1478.

Schroeder, W. A., and Holmquist, W. R. (1966). A demountable tube for large-scale chromatography and its application to the isolation of hemoglobin A$_{Ic}$. *J. Chromatog. 23:*248-253.

Schroeder, W. A., Jakway, J., and Powars, D. (1973). Detection of hemoglobins S and C at birth. A rapid screening procedure by column chromatography. *J. Lab. Clin. Med. 82:*303-308.

Schroeder, W. A., Pace, L. A., and Huisman, T. H. J. (1976). Chromatography of hemoglobins on CM-cellulose with Bis-tris and sodium chloride developers. *J. Chromatog. 118:*295-302.

Schroeder, W. A., Pace, L. A., and Huisman, T. H. J. (1978). Microchromatography of hemoglobins. VIII. A general qualitative and quantitative method in plastic drinking straws and the quantitative analysis of Hb-F. *J. Chromatog., Biomed. Applns.* *145:*203-212.

Schroeder, W. A., Cua, J. T., Matsuda, G., and Fenninger, W. D. (1962). Hemoglobin F$_I$, an acetyl-containing hemoglobin. *Biochim. Biophys. Acta 63:*532-534.

Schroeder, W. A., Shelton, J. R., Shelton, J. B., Cormick, J., and Jones, R. T. (1963). The amino acid sequence of the γ chain of human fetal hemoglobin. *Biochemistry* 2:992-1008.

Schroeder, W. A., Shelton, J. R., Shelton, J. B., Robberson, B., and Babin, D. R. (1967). A comparison of amino acid sequences in the β-chains of adult bovine hemoglobins A and B. *Arch. Biochem. Biophys.* 120:124-135.

Schroeder, W. A., Huisman, T. H. J., Shelton, J. R., and Wilson, J. B. (1970). An improved method for quantitative determination of human fetal hemoglobin. *Anal. Biochem.* 35:235-243.

Schroeder, W. A., Shelton, J. R., Shelton, J. B., Apell, G., Huisman, T. H. J., Smith, L. L., and Carr, W. R. (1972). Amino acid sequences in the β-chains of adult bovine hemoglobins C-Rhodesia and D-Zambia. *Arch. Biochem. Biophys.* 152: 222-232.

Schroeder, W. A., Huisman, T. H. J., Powars, D., Evans, L., Abraham, E. C., and Lam, H. (1975). Microchromatography of hemoglobins. IV. An improved procedure for the detection of hemoglobins S and C at birth. *J. Lab. Clin. Med.* 86:528-532.

Schroeder, W. A., Evans, L., Grussing, L., Abraham, E. C., Huisman, T. H. J., Lam, H., and Shelton, J. B. (1976). Quantitative microchromatographic determination of hemoglobin F in patients with hemoglobins S and/or C. *Am. J. Hematol.* 1:331-338.

Shukla, S. B., and Headings, V. E. (1974). Quantitation of hemoglobins by immunodiffusion. Specific antibodies to hemoglobins A_1, S, and F. *Immunochemistry* 11:741-746.

Sober, H. A., and Peterson, E. A. (1958). Protein chromatography on ion exchange cellulose. *Fed. Proc.* 17:1116-1126.

Sober, H. A., Gutter, F. J., Wyckoff, M. M., and Peterson, E. A. (1956). Chromatography of proteins. II. Fractionation of serum protein on anion-exchange cellulose. *J. Am. Chem. Soc.* 78:756-763.

Srivastava, S. K., van Loon, C., and Beutler, E. (1972). Characterization of a previously unidentified hemoglobin fraction. *Biochim. Biophys. Acta* 278:617-621.

Stamatoyannopoulos, G., Nute, P. E., Adamson, J. W., Bellingham, A. J., Funk, D., and Hornung, S. (1973). Hemoglobin Olympia

(β 20 Valine $\rightarrow$ Methionine). An electrophoretically silent variant associated with high oxygen affinity and erythrocytosis. *J. Clin. Invest.* *52*:342-349.

Steqink, L. D., Meyer, P. D., and Brummel, M. C. (1971). Human fetal hemoglobin F_1—acetylation status. *J. Biol. Chem.* *246*: 3001-3007.

Trivelli, L. A., Ranney, H. M., and Lai, H.-T. (1971). Hemoglobin components in patients with diabetes mellitus. *N. Engl. J. Med.* *284*:353-357.

VandeBerg, J. L., and Johnston, P. G. (1977). A simple technique for long-term storage of erythrocytes for enzyme electrophoresis. *Biochem. Genet.* *15*:213-215.

Vinograd, J., and Hutchinson, W. D. (1960). Carbon-14 labelled hybrids of haemoglobin. *Nature* (Lond.) *187*:216-218.

Wajcman, H., Belkhodja, O., and Labie, D. (1972). Hb Setif: Gl(94)α Asp $\rightarrow$ Tyr. A new α chain hemoglobin variant with substitution of the residue involved in a hydrogen bond between unlike subunits. *FEBS Lett.* *27*:298-300.

Wajcman, H., Labie, D., and Schapira, G. (1973). Two new hemoglobin variants with deletion. Hemoglobin Tours: Thr β87 (F3) deleted, and Hemoglobin St. Antoine: Gly-Leu β 74-75 (E 18-19) deleted. Consequences for oxygen affinity and protein stability. *Biochim. Biophys. Acta* *295*:495-504.

Waring, G., Poon, M.-C., and Bowen, S. T. (1970). The haemoglobins of *Artemia salina*. II. Isolation of three haemoglobins. *Int. J. Biochem.* *1*:537-545.

Weatherall, D. J. (ed.) (1974). Abnormal haemoglobins. *Clin. Haematol.* *3*:215-574.

Weatherall, D. J. (ed.) (1976). Haemoglobin. Structure, function, and synthesis. *Br. Med. Bull.* *32*:193-287.

Weatherall, D. J., and Clegg, J. B. (1972). *The Thalassemia Syndromes*. Blackwell, Oxford.

Weatherall, D. J., Pembrey, M. E., and Pritchard, J. (1974). Fetal haemoglobin. *Clin. Haematol.* *3*:467-508.

This appendix lists numerous hemoglobin variants that have been quantitated and/or isolated by one or more types of ion-exchange chromatography. These tables are presented only as a demonstration of the applicability of the chromatographic procedures. The method(s) named for a specific variant is (are) not necessarily the best or the only procedure(s) to be followed; on the contrary, more advanced procedures should probably be followed for several variants that have been described in earlier publications.

TABLE 1 Chromatography of Abnormal Hemoglobins, α Chain Variants

Substitution	Name	Quantitation of Variant[a]			Method used for Isolation	Ref.
		Method	% X_2	% X		
5 (A3) Ala → Asp	J-Toronto	None	–	–	DES[b]; Tris-HCl	1
6 (A4) Asp → Ala	Sawara	DES; Tris-HCL	?	17	DES; Tris-HCl	2
12 (A10) Ala → Asp	J-Paris-I	DEC[b]; Gly-KCN-NaCl	2.0	25.0	DES; Tris-HCl DEC; Gly-KCN-NaCl	3 4
15 (A13) Gly → Asp	I-Interlaken	None	–	–	AMB[b]; DEV-2 CMC[b]; Phosphate DES; Tris-HCl	5 6 7
15 (A13) Gly → Arg	Ottawa	DES; Tris-HCl	0.9	25.4	DES; Tris-HCl	8
16 (A14) Lys → Glu	I-Philadelphia	DES; Tris-HCl DEC; Gly-KCN-NaCl CMC; Phosphate	? ?	23.7	DES; Tris-HCl DEC; Gly-KCN-NaCl CMS[b]; Tris-HCl AMB; Dev-2	5 4 9 10
18 (A16) Gly → Arg	Handsworth	None			DES; Tris-HCl	11
19 (AB1) Ala → Asp	J-Kurosh	None	–	–	DES; Tris-HCl	12
21 (B2) Ala → Asp	J-Nyanza	DES; Tris-HCl	0.7	35.3	DES; Tris-HCl	13
23 (B4) Glu → Lys	Chad	DES; Tris-HCl CMC; Phosphate	? ?	16.1* 18.8*	DES; Tris-HCl (*Includes Hb A_2)	14 15
23 (B4) Glu → Gln	Memphis	None	–	–	AMB; Dev-2	16
27 (B8) Glu → Val	Spanish Town	DEC; Tris-HCl	0.7-1.0	11.5-11.6	DES; Tris-HCl	17

206

Position	Hb	Method 1			Method 2	No.
27 (B8) Glu → Gly	Fort Worth	None	–	–	DES; Tris-HCl	18
30 (B11) Glu → Lys	O-Padova	None	–	–	DES; Tris-HCl	19
30 (B11) Glu → Gln	G-Chinese	None	–	–	AMB; Dev-2	20
		DES; Tris-HCl	–	36	DES; Tris-HCl	21
45 (CD3) His → Arg	Fort De France	DES; Tris-HCl	–	20	DES; Tris-HCl	22
47 (CE5) Asp → His	Hasharon	DEC; Tris-HCl	0.6-0.8	14.2-17.7	DEC; Tris-HCl	23
					DES; Tris-HCl	24
47 (CE5) Asp → Gly	Beilinson	DEC; Gly-KCN-NaCl	1.0	33.4	DEC; Gly-KCN-NaCl	5
					AMB; Na-citrate-citric acid	25
47 (CE5) Asp → Asn	Arya	None	–	–	DEC; Tris-HCl	26
48 (CE6) Leu → Arg	Montgomery	DES; Tris-HCl	1.0	33.5	DES; Tris-HCl	27
		DEC; Gly-KCN-NaCl				4
50 (CE8) His → Asp	J-Sardegna	None	–	–	DEC; Tris-HCl	28
51 (CE9) Gly → Arg	Russ	DEC; Tris-HCl	0.4	10.7-12.3	DES; Tris-HCl	29
51 (CE9) Gly → Asp	J-Abidjan	None	–	–	DES; Tris-HCl	30
53 (E2) Ala → Asp	J-Rovigo	None	–	–	CMC; developer not stated	31
54 (E3) Gln → Glu	J-Mexico	AMB; Dev-2	?	20	AMB; Dev-2 and 6	32
		DES; Tris-HCl	–	–		33
54 (E3) Gln → Arg	Shimonoseki	AMB; Dev-2 or Na-citrate-citric acid	–	–	AMB; Dev-2 or Na-citrate-citric acid	34

TABLE 1 Chromatography of Abnormal Hemoglobins, α Chain Variants (Continued)

Substitution	Name	Quantitation of Variant[a]			Method used for Isolation	Ref.
		Method	% X_2	% X		
57 (E6) Gly → Arg	Persian Gulf	AMB; Na-citrate-citric acid	–	–	AMB; Na-citrate-citric acid	35
57 (E6) Gly → Asp	J-Norfolk	CMC; Phosphate	1.8	16	AMB; Na-citrate-citric acid	36
					CMC; Phosphate	37
58 (E7) His → Tyr	M-Boston	AMB; Dev-2	?	20	AMB; Phosphate and ammonium sulfate	38
61 (E10) Lys → Asn	J-Buda	DES; Tris-HCl (sample contained J_2, J, A_2, G-Pest, G_2)	?	21	DES; Tris-HCl	39
63 (E12) Phe → Asp	J-Pontoise	None	–	–	DES; Tris-HCl	40
64 (E13) Asp → Tyr	Perspolis	None	–	–	DEC; Tris-HCl	11
64 (E13) Asp → His	Q-India	DES; Tris-HCl	0.5	8.6	DEC; Tris-HCl	41
64 (E13) Asp → Asn	G-Waimanalo	DES; Tris-HCl	(+A_2) 3.8	17.3	DES; Tris-HCl	42
68 (E17) Asn → Lys	G-Philadelphia	DEC; Gly-KCN-NaCl	0.6-1.1	33-37	DEC; Gly-KCN-NaCl	4
		DEC; Phosphate	1.4	46.8	AMB; Na-citrate-citric acid	43
		DES; Tris-HCl	0.5-1.2	32-48	DES; Tris-HCl	44
					CMC; Phosphate	44
68 (E17) Asn → Asp	Ube II	None	–	–	DEC; Gly-KCN-NaCl	4

Position	Variant	Method	Value	Value	Method	Ref
71 (E20) Ala → Glu	J-Habana	DES; Tris-HCl	?	38	DEC; Tris-HCl	45
72 (EF1) His → Arg	Daneshgah-Tehran	None	-	-	DEC; Tris-HCl	46
74 (EF3) Asp → Asn	G-Pest	DES; Tris-HCl (see 61 J-Buda)	(+A₂) 2	23	DES; Tris-HCl	39
74 (EF3) Asp → His	Mahidol	DES; Tris-HCl	0.5-1.1	22.6-30.6	DES; Tris-HCl	47
74 (EF3) Asp → Gly	Chapel Hill	DES; Tris-HCl	0.7	23.5	DES; Tris-HCl	48
75 (EF4) Asp → Asn	Matsue-Oki	DES; Tris-HCl	1.1	32.7	DES; Tris-HCl	49
78 (EF7) Asn → Lys	Stanleyville II	None	-	-	AMB; Na-citrate-citric acid / DES; Tris-HCl	50 / 51
80 (F1) Leu → Arg	Ann Arbor	None	-	-	DES; Tris-HCl	52
85 (F6) Asp → Asn	G-Norfolk	None	-	-	DES; Tris-HCl	53
85 (F6) Asp → Tyr	Atago	CMC; Phosphate	0.7	22.8	CMC; Phosphate / DES; Tris-HCl	54 / 5
87 (F8) His → Tyr	M-Iwate	AMB; Dev-1	?	30	AMB; Dev-1 / AMB; Na-citrate-citric acid / DES; Tris-HCl	55 / 56,57 / 5
90 (FG2) Lys → Asn	Broussais	None	-	-	DES; Tris-HCl	58
90 (FG2) Lys → Thr	J-Rajappen	DES; Tris-HCl	?	25	DES; Tris-HCl	59
91 (FG3) Leu → Pro	Pórt Phillip	DES; Tris-HCl	-	7.0	DES; Tris-HCl	60
92 (FG4) Arg → Leu	Chesapeake	None	-	-	DES; Tris-HCl / DEC; Tris-HCl / CMC; Phosphate	61 / 61 / 61

TABLE 1 Chromatography of Abnormal Hemoglobins, α Chain Variants (Continued).

Substitution	Name	Quantitation of Variant[a]			Method used for Isolation	Ref.
		Method	% X$_2$	% X		
92 (FG4) Arg → Gln	J-Cape Town	None	–	27	DES; Tris-HCl	62
94 (G1) Asp → Tyr	Setif	None	–	–	DES; Tris-HCl	63
94 (G1) Asp → Asn	Titusville	DES; Tris-HCl	(+A$_2$) 3.4	34.7	DES; Tris-HCl	64
95 (G2) Pro → Ala	Denmark Hill	None	–	–	CMS; Tris-HCl	65
95 (G2) Pro → Arg	St. Luke's	DES; Tris-HCl	0.4-0.6	8.9-11.1	DES; Tris-HCl	66
		DEC; Gly-KCN-NaCl	0.5-0.6	7.8-11.1	DEC; Gly-KCN-NaCl	4
		CMS; Tris-HCl	–	14	CMS; Tris-HCl	67
95 (G2) Pro → Ser	Rampa	None	–	–	DEC; Tris-phosphate	68
					DES; Tris-HCl	69
95 (G2) Pro → Leu	G-Georgia	DES; Tris-HCl	0.7-1.0	9-11;23	DES; Tris-HCl	70
		DEC; Gly-KCN-NaCl	0.3-1.1	13.6;22.4	DEC; Gly-KCN-NaCl	4

Mutation	Name	Method	Value	Value	Method	Ref.
102 (G9) Ser → Arg	Manitoba	DES; Tris-HCl	0.6	15	DES; Tris-HCl	71
112 (G19) His → Arg	Strumica	DES; Tris-HCl	?	16-17	DES; Tris-HCl	72
114 (GH2) Pro → Arg	Chiapas	AMB; Dev-2	?	25	AMB; Dev-2	32
116 (GH2) Glu → Lys	O-Indonesia	CMC; Phosphate	?	11.6*	DES; Tris-HCl / AMB; Na-citrate-citric acid	14 / 73
		DEC; Tris-HCl (*contains Hb A$_2$)		18*	DEC; Gly-KCN-NaCl	4,74
120 (H3) Ala → Glu	J-Birmingham	DES; Tris-HCl	?	25	DES; Tris-HCl	75
126 (H9) Asp → Asn	Tarrant	DES; Tris-HCl	0.5	19.5	DES; Tris-HCl	76
127 (H10) Lys → Thr	St. Claude	DES; Tris-HCl	1.2	32.7	DES; Tris-HCl	77
127 (H10) Lys → Asn	Jackson	DES; Tris-HCl	0.9	25.3	DES; Tris-HCl	78
136 (H19) Leu → Pro	Bibba	DES; Tris-HCl	?	5.1	DES; Tris-HCl	79
141 (HC3) Arg → Gly	J-Camagüey	DES; Tris-HCl	-	18	DES; Tris-HCl	80
141 (HC3) Arg → His	Suresnes	DES; Tris-HCl / DEC; Gly-KCN-NaCl	0.65-0.8	39 / 21.8-22	DES; Tris-HCl / DEC; Gly-KCN-NaCl	81 / 82

TABLE 2 Chromatography of Abnormal Hemoglobins, β Chain Variants

Substitution	Name	Quantitation of Variant[a] Method	% X[a]	Method Used for Isolation	Ref.
1 (NA1) Val → Ac-Ala	Raleigh			DES[b]; Tris-HCl CMC[b]; Phosphate	83
2 (NA2) His → Arg	Deer Lodge	DES; Tris-HCl DEC; Gly-KCN-NaCl CMC; Gly-KCN-NaCl	44	DES[b]; Tris-HCl	84 85 85
6 (A3) Glu → Val	S	DEC; Phosphate DEC; Gly-KCN-NaCl DES; Tris-HCl CMC; Tris-NaCl-KCN CMS; Tris-HCl AMB; Dev-2 AMB; Na-citrate-citric-acid *Using various methods.	28–46*	Same as listed under analytical procedures.	86 4 87 86 88 89 89
6 (A3) Glu → Lys	C	Same as listed under Hb S. *Using various methods; often includes Hb A$_2$.	30–48*	Same as listed under Hb S.	
7 (A4) Gly → Gly	G-San Jose	AMB; Dev-2 DES; Tris-HCl	24–28	AMB[b]; Dev-2	90 91
7 (A4) Glu → Lys	C-Siriraj	DES; Tris-HCl CMC; Phosphate *Includes Hb A$_2$.	30–40* 22.3	DES; Tris-HCl CMC; Phosphate AMB; Na-citrate-citric acid	92 14 93

Position	Name	%	Analytical procedure	Preparative procedure	Ref
9 (A6) Ser → Cys	Porto Alegre		None	Sephadex G-100 in 0.2 M NaCl; see note	94
10 (A7) Ala → Asp	Ankara	42	DES; Tris-HCl	DES; Tris-HCl	95
14 (A11) Leu → Pro	Saki		DES; Tris-HCl	DES; Tris-HCL; see note	96
15 (A12) Trp → Arg	Belfast	29	DES; Tris-HCl	DES; Tris-HCl	97
16 (A13) Gly → Asp	J-Baltimore	38-40 / 36-40	DEC; Tris-HCl / DES; Tris-HCl	Same as listed under analytical procedures.	98 / 99
17 (A14) Lys → Glu	Nagasaki	43	CMC; Phosphate	CMC; Phosphate	100
19 (B1) Asn → Asp	Alamo	52-54	DEC; Gly-NaCl-KCN	DEC; Gly-NaCl-KCN	101
22 (B4) Glu → Lys	E-Saskatoon	46*	CMC; Phosphate *Contains Hb A$_2$	CMC; Phosphate	14
22 (B4) Glu → Gln	D-Iran		None	None	102
24 (B6) Gly → Arg	Riverdale-Bronx	?*	DEC; Tris-HCl	DEC; Tris-HCl	103
24 (B6) Gly → Val	Savannah	?*	DES; Tris-HCl	DES; Tris-HCl	5
24 (B6) Gly → Asp	Moscova	?*	DES; Tris-HCl *Unstable; quantity uncertain	DES; Tris-HCl	104

[Note: Separation of Hb Porto Alegre from Hb A and Hb A$_2$ is based on the special property of the variant, i.e. formation of polymers.]

[Note: The sample used was from a patient with the Hb S-Hb Saki combination. Hb Saki and Hb A cannot be separated by this type of chromatography.]

213

TABLE 2 Chromatography of Abnormal Hemoglobins, β Chain Variants (Continued)

Substitution	Name	Quantitation of Variant[a] Method	% X[a]	Method Used for Isolation	Ref.
26 (B8) Gly → Lys	E	DEC; Tris-HCl		Same as listed under	14←
		DEC; Gly-KCN-NaCl	32-45*	analytical procedures.	105
		DES; Tris-HCl	32-45*		106
		CMC; Phosphate	36*		106
		*Contains Hb A$_2$			
26 (B8) Glu → Val	Henri Mondor			DES; Tris-HCl	107
28 (B10) Leu → Pro	Genova	Several procedures tested but found unreliable because of instability of variant.		DES; Tris-HCl	108
30 (B12) Arg → Ser	Tacoma	DES; Tris-HCl	42	DES; Tris-HCl	109
37 (C3) Trp → Ser	Hirose	CMC; Phosphate	41	CMC; Phosphate	110
37 (C3) Trp → Arg	Rothschild			DES; Tris-HCl	111
39 (C5) Gln → Lys	Alabama	DES; Tris-HCl	38	DES; Tris-HCl	27
39 (C5) Gln → Glu	Vaasa	DEC; Gly-KCN-NaCl		DEC; Gly-KCN-NaCl	112
40 (C6) Arg → Lys	Athens-Georgia	DEC; Gly-KCN-NaCl	47-52	DEC; Gly-KCN-NaCl	113
				CMC; Phosphate	113
40 (C6) Arg → Ser	Austin	DES; Tris-HCl	45	DES; Tris-HCl	114

43 (CD2) Glu → Ala	G-Galveston	AMB; Na-citrate-citric acid	–	AMB; Na-citrate-citric acid	115
48 (CD7) Leu → Arg	Okaloosa	None	–	DES; Tris-HCl	116
51 (D2) Pro → Arg	Willamette	DES; Tris-HCl	32-35	DES; Tris-HCl	117
52 (D3) Asp → Asn	Osu Christians-borg	None	–	DES; Tris-HCl	118
56 (D7) Gly → Asp	J-Meinung	None	–	DES; Tris-HCl	119
56 (D7) Gly → Arg	Hamadan	None	–	DEC; Tris-HCl / AMB; different phosphate buffers	120, 5
59 (E3) Lys → Thr	J-Honolulu	DES; Tris-HCl	47	DES; Tris-HCl / CMC; Phosphate	121, 122
59 (E3) Lys → Asn	J-Lome			DES; Tris-HCl	123
60 (E4) Val → Leu	Yatsushiro	CMC; Phosphate	45	CMC; Phosphate	124
63 (E7) His → Arg	Zurich	CMC; Phosphate	28-34	CMC; Phosphate / CMS; Phosphate	125, 126
63 (E7) His → Tyr	M-Saskatoon	DES; Tris-HCl	22-31	DES; Tris-HCl / AMB; 0.05 M K-phosphate, pH 7.0, with NaCl gradient	127, 128
64 (E8) Gly → Asp	J-Calabria	None	–	DES; Tris-HCl	129
65 (E9) Lys → Asn	Sicilia	None	–	DES; Tris-HCl	130
65 (E9) Lys → Gln	J-Cairo	None	–	DES; Tris-HCl	131
66 (E10) Lys → Glu	I-Toulouse	None	–	AMB; Dev-2	132

TABLE 2 Chromatography of Abnormal Hemoglobins, β Chain Variants (Continued)

Substitution	Name	Quantitation of Variant[a] Method	% X[a]	Method Used for Isolation	Ref.
67 (E11) Val → Glu	M-Milwaukee	None	–	AMB; 0.1 M K-phosphate, pH 7.0, with NaCl gradient	133
70 (E14) Ala → Asp	Seattle	DES; Tris-HCl	39-44	DES; Tris-HCl	134
73 (E17) Asp → Val	Mobile	None	–	DES; Tris-HCl	135
73 (E17) Asp → Asn	Korle Bu	DES; Tris-HCl	35-42	DES; Tris-HCl	136
73 (E17) Asp → Tyr	Vancouver		–	DES; Tris-HCl	137
74 (E18) Gly → Val	Bushwick	None	–	DEC; Tris-phosphate	138
74 (E18) Gly → Asp	Shepherds Bush	DES; Tris-HCl *Value too low; variant unstable	24*	DES; Tris-HCl	139
76 (E20) Ala → Asp	J-Chicago	None	–	DES; Tris-HCl	140
79 (EF3) Asp → Gly	G-Hsi-Tsou	DES; Tris-HCl	46	DES; Tris-HCl	141
80 (EF4) Asn → Lys	G-Szuhu			DES; Tris-HCl	142
81 (EF5) Leu → Arg	Baylor		–	DES; Tris-HCl	143
82 (EF6) Lys → Met	Helsinki			DES; Tris-HCl	144

Position	Name	Procedure	Value	Analytical procedures	Ref.
82 (EF6) Lys → Asn / → Asp	Providence	DES; Tris-HCl (high pressure liquid chromatography)	Prov. Asn: 19 / Prov. Asn: 32	Same as listed under analytical procedures	145
82 (EF6) Lys → Thr	Rahere	DES; Tris-HCl	48		146
83 (EF7) Gly → Asp	Pyrgos	DEC; Tris-phosphate (a sample from a patient with Hb S-Hb Pyrgos) CMC; Phosphate	62 / 44.5-46.5	Same as listed under analytical procedures.	147
83 (EF7) Gly → Cys	Ta-Li	DES; Tris-HCl	40	DES; Tris-HCl	148
87 (F3) Thr → Lys	D-Ibadan	DEC; Gly-KCN-NaCl		DEC; Gly-KCN-NaCl	4
89 (F5) Ser → Asn	Creteil	DES; Tris-HCl	30	DEC; Tris-HCl	149
89 (F5) Ser → Arg	Vanderbilt			DES; Tris-HCl	150
90 (F6) Glu → Lys	Agenogi	CMC; Phosphate DES; Tris-HCl *Contains Hb A2	40-45* / 45.5	CMC; Phosphate	151,14 / 152
91 (F7) Leu → Pro	Sabine	DES; Tris-HCl	7-10	DES; Tris-HCl	153
91 (F7) Leu → Arg	Caribbean	DES; Tris-HCl	39	DES; Tris-HCl	154
92 (F8) His → Tyr	M-Hyde Park	DES; Tris-HCl	22-31*	DES; Tris-HCl AMB; Phosphate buffers	155 / 156
92 (F8) His → Gln	Istanbul	DES; Tris-HCl (0.024 M) *Value too low; variant unstable	12*	DES; Tris-HCl (0.025 M)	157
92 (F8) His → Pro	Newcastle			DES; Tris-HCl	158

TABLE 2 Chromatography of Abnormal Hemoglobins, β Chain Variants (Continued)

Substitution	Name	Quantitation of Variant[a]		Method Used for Isolation	Ref.
		Method	% X^a		
95 (FG2) Lys → Glu	N-Baltimore	DES; Tris-HCl DEC; Gly-KCN-NaCl	48-52 49-52	Same as listed under analytical procedures.	159 4
97 (FG4) His → Leu	Wood	None	–	CMS; Phosphate	160
97 (FG4) His → Gln	Malmö	DEC; Gly-KCN-NaCl DES; Tris-HCl CMC; Phosphate (interrupted gradient)	46 35 40	Same as listed under analytical procedures.	4 161 162
98 (FG5) Val → Met	Köln	DES; Tris-HCl	?	DES; Tris-HCl CMC; Phosphate CMC; Phosphate AMB; Na-citrate-citric acid	163,164 163,164 165 166
99 (G1) Asp → His	Yakima	DES; Tris-HCl	38.5	DES; Tris-HCl DEC; 0.01 M Tris-HCl; elution with a linear NaCl gradient	167 168
99 (G1) Asp → Asn	Kempsey	DES; Tris-HCl	45-47	DES; Tris-HCl DEC; 0.01 M Tris-HCl; elution with a linear NaCl gradient	169 168
99 (G1) Asp → Ala	Radcliffe			DES; Tris-HCl	170
101 (G3) Glu → Gly	Alberta	DEC; Gly-KCN-NaCl	45	DEC; Gly-KCN-NaCl	171

Position	Name	Preparative procedure	%	Analytical procedure	Ref.
101 (G3) Glu → Lys	British Columbia			DES; Tris-HCl	172
102 (G4) Asn → Thr	Kansas	CMC; Phosphate	±50	CMC; with a gradient from 0.01 M phosphate, pH 7.0, to 0.01 M phosphate, pH 7.5	173
102 (G4) Asn → Ser	Beth Israel	None	–	AMB; 0.131 M phosphate, pH 6.42	174
		DES; Tris-HCl DEC; Gly-KCN-NaCl	41	DES; Tris-HCl DEC; Gly-KCN-NaCl	175 175
102 (G4) Asn → Lys	Richmond	DES; Tris-HCl DEC; Gly-KCN-NaCl	30–38 30–40	DES; Tris-HCl DEC; Gly-KCN-NaCl	176 4
104 (G6) Arg → Ser	Camperdown	DES; Tris-HCl	50	DES; Tris-HCl	177
106 (G8) Leu → Gln	Tübingen	DES; Tris-HCl (as methemoglobin only)	41	DES; Tris-HCl	178
108 (G10) Asn → Asp	Yoshizuka	DES; Tris-HCl	50	DES; Tris-HCl	179
111 (G13) Val → Phe	Peterborough	DES; Tris-HCl	33.5	DES; Tris-HCl	180
113 (G15) Val → Glu	New York	None	–	AMB; Dev-2	181
117 (G19) His → Arg	P-Galveston	None	–	DES; Tris-HCl AMB; Na-citrate-citric acid	182 183
119 (GH2) Gly → Asp	Fannin-Lubbock	DES; Tris-HCl	44.5	DES; Tris-HCl	184,185
120 (GH3) Lys → Asn	Riyadh	DES; Tris-HCl	–	DES; Tris-HCl	186
121 (GH4) Glu → Lys	O-Arab	CMC; Phosphate DES; Tris-HCl DEC; Gly-KCN-NaCl *Contains Hb A_2	38 44* 40–44*	Same as listed under analytical procedures.	14 187 4

TABLE 2 Chromatography of Abnormal Hemoglobins, β Chain Variants (Continued)

Substitution	Name	Quantitation of Variant[a]		Method Used for Isolation	Ref.
		Method	% X[a]		
121 (GH4) Glu → Gln	D-Los Angeles	Same as listed under O-Arab	30–40	See O-Arab	
121 (GH4) Glu → Val	Beograd	Same as listed under O-Arab	30–40	See O-Arab	
127 (H5) Gln → Glu	Haceteppe	DES; Tris–HCl	45.5	DES; Tris–HCl	188
128 (H6) Ala → Asp	J-Guantanamo	DES; Tris–HCl	36–38	DES; Tris–HCl	189
129 (H7) Ala → Asp	J-Taichung	DES; Tris–HCl	?	DES; Tris–HCl	190
131 (H9) Gln → Glu	Camden	DES; Tris–HCl	46.5	Same as listed under analytical procedures.	191,192
		DEC; Gly-KCN-NaCl	45–47		4
		CMC; Phosphate			192

Position	Name	Analytical procedure	%	Preparative procedure	Ref.
132 (H10) Lys → Gln	K-Woolwich	DES; Tris-HCl	?	DES; Tris-HCl	193
136 (H14) Gly → Asp	Hope	DES; Tris-HCl DEC; Gly-KCN-NaCl	45-50 49.5	DES; Tris-HCl DEC; Gly-KCN-NaCl	194,192 4
143 (H21) His → Gln	Little Rock	None	–	CMC; Phosphate	195
143 (H21) His → Pro	Syracuse	None	–	CMC; Phosphate	196
144 (HC1) Lys → Asn	Andrew-Minneapolis	None	–	DES; Tris-HCl	197
145 (HC2) Tyr → Asp	Fort Gordon	DES; Tris-HCl CMC; Phosphate	28 30	DES; Tris-HCl CMC; Phosphate	198 198
145 (HC2) Tyr → Cys	Rainier	CMS; 0.05 M phosphate, pH 6.2, with 0.1 M NaCl; gradient to 0.2 M NaCl	30	Same as listed under analytical procedures.	199
145 (HC2) Tyr → His	Bethesda	Estimated from data obtained in preparative chromatography	46	AMB; 0.1 M phosphate, pH 7.0; variant is eluted with 0.1 M K phosphate + 0.5 M NaCl, pH 7.0 DES; Tris-HCl CMC; Phosphate	200 200 201
146 (HC3) His → Arg	Port Royal	DES; Tris-HCl	48	DES; Tris-HCl	202

TABLE 3   Chromatography of Abnormal Hemoglobins, δ Chain Variants

Substitution	Name	Quantitation of Variant[a]		Method Used for Isolation	Ref.
		Method	% X_2		
2 (NA2) His → Arg	A₂-Sphakiá	DES; Tris-HCl	0.6	DES[b]; Tris-HCl	203
12 (A9) Asn → Lys	A₂-NYU	DEC; Phosphate DES; Tris-HCl CMC; Phosphate	– 0.76 0.5-1.5	DES; Tris-HCl DEC[b]; Phosphate DES; Tris-HCl CMC[b]; Phosphate	204 205 206 206
16 (A13) Gly → Arg	A₂′ (B2)	DES; Tris-HCl CMC; Phosphate	– 0.5-1.5	DEC; Tris-HCl DES; Tris-HCl CMC; Phosphate	207 206 206
20 (B2) Val → Glu	A₂-Roosevelt	DEC; Tris-phosphate	0.5	DEC; Tris-phosphate	208
22 (B4) Ala → Glu	A₂-Flatbush	DES; Tris-HCl CMC; Phosphate	– 0.5-1.5	DEC; Tris-HCl DES; Tris-HCl CMC; Phosphate	207 206 206
43 (CD2) Glu → Lys	A₂-Melbourne	DES; Tris-HCl	1.2	DES; Tris-HCl	209
51 (D2) Pro → Arg	A₂-Adria	DES; Tris-HCl	1.6-2.6	DES; Tris-HCl	210
69 (E13) Gly → Arg	A₂-Indonesia	DES; Tris-HCl	1.0-2.6	DES; Tris-HCl	211
116 (G18) Arg → His	A₂-Coburg	DES; Tris-HCl	1.4-2.8	DES; Tris-HCl	212
136 (H14) Gly → Asp	A₂-Babinga	DEC; Tris-phosphate DES; Tris-HCl	0.8-1.3 1.05-1.15	DEC; Tris-phosphate DES; Tris-HCl	213 214

TABLE 4 Chromatography of Abnormal Hemoglobins, γ Chain Variants

Substitution	Name	Quantitation of Variant[a]		Method Used for Isolation	Ref.
		Method	% X_2		
1 (NA1) Gly → Cys (136 Gly)	F-Malaysia	DES; Tris-HCl	14.3	DES[b]; Tris-HCl	215
5 (A2) Glu → Lys (136 Ala)	F-Texas-I	DES; Tris-HCl	6.0-9.9	DES[b]; Tris-HCl DEC[b]; Phosphate AMB[b]; Citrate	216 217 217
7 (A4) Asp → Asn (136 Gly)	F-Auckland	DES; Tris-HCl	13	DES; Tris-HCl	218
16 (A13) Gly → Arg (136 Gly)	F-Melbourne	DES; Tris-HCl	28-30	DES; Tris-HCl	219
22 (B4) Asp → Gly (136 Ala)	F-Kuala Lumpur		–	DES; Tris-HCl	220
61 (E5) Lys → Glu (136 Ala)	F-Jamaica	DES; Tris-HCl	7.0-9.0	DES; Tris-HCl AMB; Dev-6	221 221
80 (EF4) Asp → Tyr (136 Ala)	F-Victoria Jubilee	DEC; Tris-HCl CMS; Tris-HCl	4.9-6.1	DEC; Tris-HCl CMS; Tris-HCl	222 222
117 (G19) His → Arg (136 Gly)	F-Malta-I	DEC; Gly-KCN-NaCl	19.1-24.7	CMC; Phosphate DEC; Gly-KCN-NaCl	223 224
121 (GH4) Glu → Lys (136 Ala)	F-Hull	DES; Tris-HCl CMS; Tris-HCl	6.7 6.9	DES; Tris-HCl CMS; Tris-HCl	225 225
121 (GH4) Glu → Lys (136 Gly)	F-Carlton	DES; Tris-HCl	26	DES; Tris-HCl	219
125 (H3) Glu → Ala (136 Gly)	F-Port Royal	DES; Tris-HCl	12.0	DES; Tris-HCl	226

TABLE 5 Chromatography of Abnormal Hemoglobins, Special Variants

Substitution	Name	Quantitation of Variant Method	% X[a]	Method Used for Isolation	Ref.
α 141, with additional residues; residue is Gln.	Hb Constant Spring	DES; Tris-HCl	1	DES[b]; Tris-HCl AMB[b]	227 228
α 141, with 31 additional residues; residue 142 is Lys.	Hb Icaria	None	?	AMB. The method is about the same as for Hb CS. Icaria is more tightly bound to the resin than is Hb CS.	229
α 141, with likely 31 additional residues; residue 142 is Ser.	Hb Koya Dora	None	?	DEC[b]	230
α 139-141, with 5 additional residues. (Frame shift)	Hb Wayne	DES; Tris-HCl	2-3	DES; Tris-HCl	231
β 145-146, with 8 additional residues. (Frame shift)	Hb Cranston	DEC; Tris-HCl	35	DEC; Tris-HCl	232
α; residues 116, 117 and 118 (Glu-Phe-Thr) repeated (Insertion)	Hb Grady	DEC; Tris-HCl DES; Tris-HCl	8-18	DEC; Tris-HCl DES; Tris-HCl	233 233

224

Structural feature	Hb	Analytical procedure	Value	Preparative procedure	Ref.
δ-β; crossover between residues 87δ (F3) and 116β (G18)	Hb Lepore-Washington	DES; Tris-HCl DEC; Gly-KCN-NaCl CMS^b; Tris-HCl	8-12 8-12 8-12	Same as analytical procedures	234 4 234
δ-β; crossover between residues 50δ (D1) and 86β (F2)	Hb Lepore-Baltimore	DES; Tris-HCl DEC; Gly-KCN-NaCl	7-10 8-11	Same as analytical procedures	235 4
δ-β; crossover between residues 22δ (B4) and 50β (D1)	Hb Lepore-Hollandia	DEC; Gly-KCN-NaCl (one patient)	9.5	DEC; Gly-KCN-NaCl CMC^b; Phosphate	4 236
β-δ; crossover between residues 12β (A9) and 22δ (B4)	Hb Miyada	DES; Tris-HCl CMC; Phosphate *Includes Hb A_2	15.6* 13.1*	Same as analytical procedures	237 14
β-δ; crossover between residues 22β (B4) and 50δ (D1)	Hb P-Nilotic	DES; Tris-HCl	16-18	?	238
β-δ; crossover between residues 22β (B4) and 50δ (D1)	Hb Lincoln Park		–	DES; Tris-HCl	239
γ-β; crossover between residues 81γ (EF5) and 86β (F2)	Hb Kenya	DES; Tris-HCl CMC; Phosphate DEC; Gly-KCN-NaCl *Includes Hb A_2	5.7 8.4* 5.5	Same as analytical procedures	240 241 4
residue 6β or 7β deleted	Hb Leiden	DEC; Tris-phosphate-NaCl	23-24	DEC; Tris-phosphate-NaCl	242
residue 17β and 18β (Lys-Val) deleted	Hb Lyon	None		AMB; Dev-2	243
residue 23β (Val) deleted	Hb Freiburg	DES; Tris-HCl	25	DES; Tris-HCl CMC; Phosphate	244 244

TABLE 5 Chromatography of Abnormal Hemoglobins, Special Variants (Continued)

Substitution	Name	Quantitation of Variant		Method Used for Isolation	Ref.
		Method	% X^a		
résidue 87β (Thr) deleted	Hb Tours	None		AMB; 0.05 M K-phosphate, pH 7.0, followed by the same developer + 0.3 M NaCl	245
residues 91-95β (Leu-His-Cys-Asp-Lys) deleted	Hb Gun Hill	DEC; Gly-KCN-NaCl *Includes Hb A_2	22-24*	DEC; special gradient DEC; Gly-KCN-NaCl	246 247
residue 131β (Gln) deleted	Hb Leslie	DES; Tris-HCl DEC; Gly-KCN-NaCl CMC; Phosphate	11-35	Same as analytical procedures	248 4 248
6β (A3) Glu → Val and 73β (E17) Asp → Asn	Hb C-Harlem	DES; Tris-HCl DEC; Gly-KCN-NaCl CMC; Phosphate *Includes Hb A_2	43.5* 41-43* 23.7	Same as analytical procedures	249 4 14
residue 145β (HC2) → term	Hb McKees Rock			CMS; 0.05 M Tris-Maleic acid, pH 6.7-7.0	250

aThe percentages are for heterozygotes; bDES = DEAE-Sephadex; AMB = Amberlite IRC-50; DEC = DEAE-Cellulose; CMS = CM-Sephadex; CMC = CM-Cellulose.

APPENDIX REFERENCES

1. F. Vella, J. R. Hill, B. Wiltshire, and H. Lehmann. *Clin. Biochem.* *4*:137-140 (1971).

2. I. Sumida, Y. Ohta, T. Imamura, and T. Yanase. *Biochim. Biophys. Acta 322*:23-26 (1973).

3. G. A. Niazi, G. D. Efremov, N. Nikolov, E. Hunter, Jr., and T. H. J. Huisman. *Biochim. Biophys. Acta 412*:181-186 (1975).

4. E. C. Abraham, A. Reese, M. Stallings, and T. H. J. Huisman. *Hemoglobin 1*:27-44 (1976-77).

5. T. H. J. Huisman and collaborators, unpublished observations.

6. V. H. R. Marti, C. Pik, and P. Mosimann. *Acta Haemat. 32*:9-16 (1964).

7. G. Schilirò, S. Musumeci, G. Pizzarelli, A. Russo, M. Marinucci, E. Bruni, and G. Russo. *Blood 48*:639-644 (1976).

8. F. Vella, R. Casey, H. Lehmann, A. Labossiere, and T. G. Jones. *Biochim. Biophys. Acta 336*:25-29 (1974).

9. E. W. Baur. *Humangenetik 6*:368-372 (1968).

10. C. Boulard, A. Cosset, F. Destaing, A. Duzer, J. H. P. Jonxis, C. J. Muller, and A. Portier. *Blood 18*:750-757 (1961).

11. K. D. Griffiths, A. Lang, H. Lehmann, J. R. Mann, D. Plowman, and D. N. Raine. *FEBS Letters 75*:93-95 (1977).

12. S. Rahbar, F. Ala, E. Akhavan, G. Nowzari, I. Shoa'i, and M. H. Zamanianpoor. *Biochim. Biophys. Acta 427*:119-125 (1976).

13. A. G. Kendall, R. D. Barr, A. Lang, and H. Lehmann. *Biochim. Biophys. Acta 310*:357-359 (1973).

14. S. H. Boyer, E. F. Crosby, G. F. Fuller, L. Ulenurm., and A. A. Buck. *Am. J. Hum. Genet. 20*:570-578 (1968).

15. T. H. J. Huisman and R. N. Wrightstone. *J. Chromatog. 92*:391-399 (1974).

16. L. M. Kraus, T. Miyaji, I. Iuchi, and A. P. Kraus. *Biochemistry 5*:3701-3708 (1966).

17. E. Ahern, V. Ahern, W. Holder, E. Palomino, G. R. Serjeant, B. E. Serjeant, M. Forbes, B. Brimhall, and R. T. Jones. *Biochim. Biophys. Acta 427*:530-538 (1976).

18. R. G. Schneider, B. Brimhall, R. T. Jones, R. Bryant, C. B. Mitchell, and A. I. Goldberg. *Biochim. Biophys. Acta 243*:164-169 (1971).

19. L. Vettore, G. De Sandre, E. E. Di Iorio, K. H. Winterhalter, A. Lang, and H. Lehmann. *Blood 44*:869-877 (1974).

20. R. T. Swenson, R. L. Hill, H. Lehmann, and R. T. S. Jim. *J. Biol. Chem. 237*:1517-1520 (1962).

21. R. Q. Blackwell, M.-I. Weng, C.-S. Liu, T.-B. Shih, and C.-L. Wang. *Vox Sang. 23*:363-368 (1972).

22. F. Braconnier, G. Gacon, J. Thillet, H. Wajcman, J. Soria, P. Maigret, D. Labie, and J. Rosa. *Biochim. Biophys. Acta 493*:228-233 (1977).

23. R. G. Schneider, S. Ueda, J. B. Alperin, B. Brimhall, and R. T. Jones. *Am. J. Hum. Genet. 20*:151-156 (1968).

24. S. Charache, A. M. Mondzac, and U. Gessner. *J. Clin. Invest. 48*:834-847 (1969).

25. A. De Vries, H. Joshua, H. Lehmann, R. L. Hill, and R. E. Fellows. *Br. J. Haematol. 9*:484-486 (1963).

26. S. Rahbar, N. Mahdavi, G. Nowzari, and I. Mostafavi. *Biochim. Biophys. Acta 386*:525-529 (1975).

27. B. Brimhall, R. T. Jones, R. G. Schneider, T. S. Hosty, G. Tomlin, and R. Atkins. *Biochim. Biophys. Acta 379*:28-32 (1975).

28. E. Gallo, L. Pugliatti, G. Ricco, P. G. Pich, G. Pinna, and U. Mazza. *Acta Haemat. 47*:311-320 (1972).

29. C. A. Reynolds and T. H. J. Huisman. *Biochim. Biophys. Acta 130*:541-543 (1966).

30. R. Cabannes, R. Renaud, A. Mauran, H. Pennors, D. Charlesworth, B. G. Price, and H. Lehmann. *Nouv. Rev. Fr. Hémat. 12*:289-300 (1972).

31. R. Alberti, G. M. Mariuzzi, L. Artibani, E. Bruni, and L. Tentori. *Biochim. Biophys. Acta 342*:1-4 (1974).

32. R. T. Jones, B. Brimhall, and R. Lisker. *Biochim. Biophys. Acta 154*:488-495 (1968).

33. G. Trabuchet, M. Dahmane, J. Pagnier, D. Labie, and M. Benabadji. *FEBS Letters 61*:156-158 (1976).

34. T. Miyaji, I. Iuchi, I. Takeda, and S. Shibata. *Acta Haem. Jap. 26*:531-537 (1963).

35. S. Rahbar, J. L. Kinderlehrer, and H. Lehmann. *Acta Haemat. 42*:169-175 (1969).

36. J. A. M. Ager, H. Lehmann, and F. Vella. *Br. Med. J. 2*:539-541 (1958).

37. C. Baglioni. *J. Biol. Chem. 237*:69-74 (1962).

38. A. Shimizu, A. Hayashi, Y. Yamamura, A. Tsugita, and K. Kitayama. *Biochim. Biophys. Acta 97*:472-482 (1965).

39. S. R. Hollán, J. G. Szelenyi, B. Brimhall, M. Duerst, R. T. Jones, R. D. Koler, and Z. Stocklen. *Nature 235*:47-50 (1972).

40. J. Thillet, Y. Blouquit, F. Perrone, and J. Rosa. *Biochim. Biophys. Acta 491*:16-22 (1977).

41. R. M. Schmidt, M. A. M. Ali, K. C. Bechtel, and W. F. Moo-Penn. *Am. J. Clin. Path. 66*:446-448 (1976).

42. R. Q. Blackwell, R. T. S. Jim, T. G. H. Tan, M.-I. Weng, C.-S. Liu, and C.-L. Wang. *Biochim. Biophys. Acta 322*:27-33 (1973).

43. R. G. Schneider and M. E. Haggard. *J. Lab. Clin. Med. 55*:60-66 (1960).

44. P. F. Milner and T. H. J. Huisman. *Br. J. Haematol. 34*:207-220 (1976).

45. B. Colombo, H. Vidal, H. Kamuzora, and H. Lehmann. *Biochim. Biophys. Acta 351*:1-6 (1974).

46. S. Rahbar, G. Nowzari, and P. Daneshmand. *Nature New Biol. 245*:268-269 (1973).

47. P. A. Lorkin, D. Charlesworth, H. Lehmann, S. Rahbar, S. Tuchinda, and L. E. Lie-Injo. *Br. J. Haematol. 19*:117-125 (1970).

48. E. P. Orringer, J. B. Wilson, and T. H. J. Huisman. *FEBS Letters 65*:297-300 (1976).

49. W. F. Moo-Penn, M. H. Johnson, and B. L. Therrell, Jr. *Hemoglobin 2*:71-74 (1978).

50. G. Van Ros, D. Beale, and H. Lehmann. *Br. Med. J. 4*:92-93 (1968).

51. M. L. North, P. D. Darbre, H. Lehmann, and J. G. Juif. *Acta Haemat. 53*:56-59 (1975).

52. J. G. Adams, III, W. P. Winter, D. L. Rucknagel, and H. H. Spencer. *Science 176*:1427-1429 (1972).

53. M. Cohen-Solal, B. Manesse, J. Thillet, and J. Rosa. *FEBS Letters 50*:163-167 (1975).

54. N. Fujiwara, T. Maekawa, and G. Matsuda. *Int. J. Prot. Res. 3*:35-39 (1971).

55. G. Kikuchi, N. Hayashi, and A. Tamura. *Biochim. Biophys. Acta 90*:199-201 (1964).

56. S. Shibata, A. Tamura, I. Iuchi, and H. Takahashi. *Acta Haem. Jap. 23*:96-105 (1960).

57. R. T. Jones, R. D. Coleman, and P. Heller. *J. Biol. Chem. 241*:2137-2143 (1966).

58. F. Vella, D. Charlesworth, P. A. Lorkin, and H. Lehmann. *Can. J. Biochem. 48*:908-910 (1970).

59. R. D. Hyde, J. L. Kinderlehrer, H. Lehmann, and M. D. Hall. *Biochim. Biophys. Acta 243*:515-519 (1971).

60. S. O. Brennan, G. P. Tauro, W. Melrose, and R. W. Carrell. *FEBS Letters 81*:115-117 (1977).

61. K. Imai. *J. Biol. Chem. 249*:7607-7612 (1974).

62. P. G. Gacon, K. E. Amegnizin, O. Belkhodja, P. Denis, R. Krishnamoorthy, D. Labie, R. Lefrancois, Y. Michel, P. Pasquis, and H. Wajcman. *Nouv. Rev. Fr. Hémat. 18*:35-44 (1977).

63. J. P. Aubert, F. Drupt, J. Rousseaux, and M.-H. Loucheux-Lefebvre. *FEBS Letters 84*:375-378 (1977).

64. R. G. Schneider, R. J. Atkins, T. S. Hosty, G. Tomlin, R. Casey, H. Lehmann, P. A. Lorkin, and K. Nagai. *Biochim. Biophys. Acta 400*:365-373 (1975).

65. B. G. Wiltshire, K. G. A. Clark, P. A. Lorkin, and H. Lehmann. *Biochim. Biophys. Acta 278*:459-464 (1972).

66. W. H. Bannister, J. L. Grech, C. F. Plese, L. L. Smith, B. P. Barton, J. B. Wilson, C. A. Reynolds, and T. H. J. Huisman. *Eur. J. Biochem. 29*:301-307 (1972).

67. P. A. Lorkin, R. Casey, K. G. A. Clark, and H. Lehmann. *FEBS Letters 39*:111-114 (1974).

68. W. W. W. De Jong, L. F. Bernini, and P. M. Khan. *Biochim. Biophys. Acta 236*:197-200 (1971).

69. L. L. Smith, C. F. Plese, B. P. Barton, S. Charache, J. B. Wilson, and T. H. J. Huisman. *J. Biol. Chem. 247*:1433-1439 (1972).

70. R. Wrightstone, M. Hubbard, and T. H. J. Huisman. *Acta Haemat. 51*:315-320 (1974).

71. R. N. Wrightstone, L. L. Smith, J. B. Wilson, F. Vella, and T. H. J. Huisman. *Biochim. Biophys. Acta 412*:283-287 (1975).

72. G. A. Niazi, G. D. Efremov, N. Nikolov, E. Hunter, Jr., and T. H. J. Huisman. *Biochim. Biophys. Acta 412*:181-186 (1975).

73. L. E. Lie-Injo. *Acta Haemat. 25*:368-371 (1961).

74. S. Rahbar, G. Nowzari, and M. Poosti. *J. Clin. Path. 64*:416-420 (1975).

75. R. Q. Blackwell, W. H. Boon, C.-L. Wang, M.-I. Weng, and C.-S. Liu. *Biochim. Biophys. Acta 351*:7-12 (1974).

76. W. F. Moo-Penn, D. L. Jue, M. H. Johnson, S. M. Wilson, B. Therell, Jr., and R. M. Schmidt. *Biochim. Biophys. Acta 490*:443-451 (1977).

77. F. Vella, P. Galbraith, J. B. Wilson, S. C. Wong, G. C. Folger, and T. H. J. Huisman. *Biochim. Biophys. Acta 365*:318-322 (1974).

78. W. F. Moo-Penn, K. C. Bechtel, M. H. Johnson, D. L. Jue., S. Holland, C. Huff, and R. M. Schmidt. *Am. J. Clin. Path. 66*:453-456 (1975).

79. E. F. Kleihauer, C. A. Reynolds, A. M. Dozy, J. B. Wilson, R. R. Moores, M. P. Berenson, C.-S. Wright, and T. H. J. Huisman. *Biochim. Biophys. Acta 154*:220-222 (1968).

80. G. Martinez, F. Lima, C. Residenti, and B. Colombo. *Hemoglobin* 2:47-52 (1978).

81. C. Poyart, R. Krishnamoorthy, E. Bursaux, G. Gacon, and D. Labie. *FEBS Letters* 69:103-107 (1976).

82. M. E. Gravely, H. F. Harris, M. Stallings, H. Lam, J. B. Wilson, and T. H. J. Huisman. *Hemoglobin* 2:187-189 (1978).

83. W. F. Moo-Penn, K. C. Bechtel, R. M. Schmidt, M. H. Johnson, D. L. Jue, D. E. Schmidt, Jr., W. M. Dunlap, S. J. Opella, J. Bonaventura, and C. Bonaventura. *Biochemistry* 16:4872-4879 (1977).

84. D. Powars, W. A. Schroeder, J. R. Shelton, L. Evans, and R. Vinetz. *Hemoglobin* 1:97-102 (1976-77).

85. E. C. Abraham, T. H. J. Huisman, W. A. Schroeder, L. A. Pace, and L. Grussing. *J. Chromatog.* 143:57-63 (1977).

86. W. A. Schroeder, T. H. J. Huisman, D. Powars, L. Evans, E. C. Abraham, and H. Lam. *J. Lab. Clin. Med.* 86:528-532 (1975).

87. R. N. Wrightstone, T. H. J. Huisman, and A. van der Sar. *Clin. Chim. Acta* 22:593-601 (1968).

88. A. M. Dozy and T. H. J. Huisman. *J. Chromatog.* 40:62-70 (1969).

89. H. K. Prins. *J. Chromatog.* 2:445-486 (1959).

90. R. L. Hill, R. T. Swenson, and H. C. Schwartz. *Blood* 19:573-586 (1962).

91. G. Ricco, E. Gallo, P. G. Pich, G. Rossi, R. Miniero, and U. Mazza. *Acta Haemat.* 52:180-188 (1974).

92. R. Q. Blackwell, C.-S. Liu, and C.-L. Wang. *Vox Sang.* 23:433-438 (1972).

93. S. Tuchinda, D. Beale, and H. Lehmann. *Br. Med. J.* 1:1583-1585 (1965).

94. J. Bonaventura and A. Riggs. *Science* 158:800-802 (1967).

95. A. Arcasoy, R. Casey, H. Lehmann, A. O. Cavdar, and A. Berki. *FEBS Letters* 42:121-123 (1974).

96. Y. Beuzard, P. Basset, F. Braconnier, H. E. Gammal, L. Martin, J. L. Oudard, and J. Thillet. *Biochim. Biophys. Acta* 393:182-187 (1975).

97. G. Gacon, H. Wajcman, D. Labie, B. Varet, and B. Christoforov. *Acta Haemat.* 55:313-319 (1976).

98. L. E. Lie-Injo, H. H. Fudenberg, H. Lehmann, and E. Gallo. *Haematologia* 2:9-17 (1968).

99. S. C. Wong, N. Bouver, J. B. Wilson, and T. H. J. Huisman. *Clin. Chim. Acta* 35:521-522 (1971).

100. M. Maekawa, T. Maekawa, N. Fujiwara, K. Tabara, and G. Matsuda.
      Int. J. Prot. Res. 2:147-156 (1970).

101. H. Lam, J. B. Wilson, H. Harris, M. Gravely, and T. H. J.
      Huisman. *Hemoglobin 1*:703-706 (1977).

102. S. Rahbar. *Br. J. Haematol. 24*:31-35 (1973).

103. H. M. Ranney, A. S. Jacobs, L. Udem, and R. Zalusky. *Biochem.
      Biophys. Res. Commun. 33*:1004-1011 (1968).

104. L. I. Idelson, N. A. Didkowsky, R. Casey, P. A. Lorkin, and
      H. Lehmann. *Nature 249*:768-770 (1974).

105. C. Altay, G. A. Niazi, and T. H. J. Huisman. *Hemoglobin 1*:
      100-102 (1976-77).

106. H. F. Bunn, W. D. Meriwether, S. P. Balcerzak, and D. L.
      Rucknagel. *J. Clin. Invest. 51*:2984-2987 (1972).

107. Y. Blouquit, N. Arous, P. E. A. Machado, M. C. Garel, and
      F. Perrone. *FEBS Letters 72*:5-7 (1976).

108. R. N. Wrightstone, J. B. Wilson, C. A. Reynolds, T. H. J.
      Huisman, S. Padmanabh, and F. Vella. *Clin. Chim. Acta 44*:
      217-227 (1973).

109. L. I. Idelson, N. A. Didkowsky, R. Casey, P. A. Lorkin, and
      H. Lehmann. *Acta Haemat. 52*:303-311 (1974).

110. K. Yamaoka. *Blood 38*:730-738 (1971).

111. G. Gacon, O. Belkhodja, H. Wajcman, D. Labie, and A. Najman.
      FEBS Letters 82:243-246 (1977).

112. A. G. Kendall, A. T. Pas, J. B. Wilson, N. Cope, K. Bolch,
      and T. H. J. Huisman. *Hemoglobin 1*:292-295 (1976-77).

113. W. J. Brown, G. A. Niazi, M. Jayalakshmi, E. C. Abraham, and
      T. H. J. Huisman. *Biochim. Biophys. Acta 439*:70-76 (1976).

114. W. F. Moo-Penn, M. H. Johnson, K. C. Bechtel, D. L. Jue,
      B. L. Therrell, Jr., and R. M. Schmidt. *Arch. Biochem.
      Biophys. 179*:86-94 (1977).

115. R. G. Schneider and M. E. Haggard. *J. Lab. Clin. Med. 55*:60-
      66 (1960).

116. S. Charache, B. Brimhall, and P. Milner. *J. Clin. Invest. 52*:
      2858-2864 (1973).

117. R. T. Jones, R. D. Koler, M. L. Duerst, and D. S. Dhindsa.
      Hemoglobin 1:45-57 (1976-77).

118. F. I. D. Konotey-Ahulu, J. L. Kinderlehrer, H. Lehmann, and
      B. Ringelhann. *J. Med. Genet. 8*:302-305 (1971).

119. U. Gunay, C. Pauli, M. Shamsuddin, R. G. Mason, W. J. Heinze,
      and G. R. Honig. *Blood 44*:683-690 (1974).

120.   S. Rahbar, G. Nowzari, H. Haydari, and P. Daneshmand.   *Bio-chim. Biophys. Acta 379*:645-648 (1975).

121.   R. Q. Blackwell, C.-S. Liu, and T.-B. Shih.   *Biochim. Biophys. Acta 229*:343-348 (1971).

122.   R. T. S. Jim and M. T. Yarbro.   *Blood 15*:285-287 (1960).

123.   H. Wajcman, K. P. E. Amegnizin, O. Belkhodja, and D. Labie.   *FEBS Letters 84*:372-374 (1977).

124.   T. Kagimoto, Y. Morino, and S. Kishimoto.   *Biochim. Biophys. Acta 532*:195-198 (1978).

125.   F. Bachmann and H. R. Marti.   *Blood 20*:272-286 (1962).

126.   K. H. Winterhalter, N. M. Anderson, G. Amiconi, E. Antonini, and M. Brunori.   *Eur. J. Biochem. 11*:435-440 (1969).

127.   G. D. Efremov, T. H. J. Huisman, M. Stanulovic, M. Zurovec, H. Duma, J. B. Wilson, and V. Jeremic.   *Scand. J. Haemat. 13*: 48-60 (1974).

128.   A. Hayashi, A. Shimizu, T. Suzuki, and Y. Yamamura.   *Biochim. Biophys. Acta 140*:251-257 (1967).

129.   Y. Blouquit, J. Thillet, Y. Beuzard, J. P. Vernant, and B. Dreyfus.   *Biochim. Biophys. Acta 492*:426-432 (1977).

130.   G. Ricco, P. G. Pich, U. Mazza, G. Rossi, F. Ajmar, P. Arese, and E. Gallo.   *FEBS Letters 39*:200-204 (1974).

131.   M. C. Garel, W. Hassan, M. T. Coquelet, M. Goosens, and J. Rosa.   *Biochim. Biophys. Acta 420*:97-104 (1976).

132.   D. Labie, J. Rosa, O. Belkhodja, and R. Bierme.   *Biochim. Biophys. Acta 236*:201-207 (1971).

133.   A. Hayashi, T. Suzuki, K. Imai, H. Morimoto, and H. Watari.   *Biochim. Biophys. Acta 194*:6-15 (1969).

134.   E. R. Huehns, F. Hecht, A. Yoshida, G. Stamatoyannopoulos, J. Hartman, and A. G. Motulsky.   *Blood 36*:209-218 (1970).

135.   R. G. Schneider, T. S. Hosty, G. Tomlin, R. Atkins, B. Brimhall, and R. T. Jones.   *Biochem. Genet. 13*:411-415 (1975).

136.   A. Miller and T. H. J. Huisman, unpublished observations.

137.   R. T. Jones, B. Brimhall, S. Pootrakul, and G. Gray.   *J. Mol. Evol. 9*:37-44 (1976).

138.   R. F. Rieder, D. J. Wolf, J. B. Clegg, and S. L. Lee.   *Nature 254*:725-727 (1975).

139.   J. M. White, M. C. Brain, P. A. Lorkin, H. Lehmann, and M. Smith.   *Nature 225*:939-941 (1970).

140.   P. L. Romain, A. D. Schwartz, M. Shamsuddin, J. G. Adams, III, R. G. Mason, L. N. Vida, and G. R. Honig.   *Blood 45*:387-393 (1975).

141. R. Q. Blackwell, T.-B. Shih, C.-L. Wang, and C.-S. Liu.
     Biochim. Biophys. Acta 257:49-53 (1972).

142. S. G. Welch. *Humangenetik 28*:331-333 (1975).

143. R. G. Schneider, R. A. Hettig, M. Bilunos, and B. Brimhall.
     Hemoglobin 1:85-96 (1976).

144. E. Ikkala, J. Koskela, R. Pikkarainen, E. L. Rahiala, M. A. F.
     El-Hazmi, K. Nagai, A. Lang, and H. Lehmann. *Acta Haemat.
     56*:257-275 (1976).

145. W. F. Moo-Penn, D. L. Jue, K. C. Bechtel, M. H. Johnson,
     R. M. Schmidt, P. R. McCurdy, J. Fox, J. Bonaventura,
     B. Sullivan, and C. Bonaventura. *J. Biol. Chem. 251*:7557-
     7562 (1976).

146. P. A. Lorkin, A. D. Stephens, M. E. J. Beard, P. F. M. Wrigley,
     L. Adams, and H. Lehmann. *Br. Med. J. 341*:200-202 (1975).

147. B. Tatsis, K. Sofroniadou, and C. I. Stergiopoulos. *Blood
     47*:827-832 (1976).

148. R. Q. Blackwell, C.-S. Liu, and C.-L. Wang. *Biochim. Biophys.
     Acta 243*:467-474 (1971).

149. J. Thillet, Y. Blouquit, M. C. Garel, B. Dreyfus, F. Reyes,
     M. Cohen-Solal, Y. Beuzard, and J. Rosa. *J. Mol. Med. 1*:
     135-150 (1976).

150. D. Puett, N. V. Paniker, K. D. Lin, J. M. Flexner, B. K.
     Wasserman, and S. B. Krantz. *Clin. Res. 25*:53A (1977).

151. T. Miyaji, H. Suzuki, Y. Ohba, and S. Shibata. *Clin. Chim.
     Acta 14*:624-629 (1966).

152. T. H. J. Huisman and collaborators, unpublished observations.

153. R. G. Schneider, S. Ueda, J. B. Alperin, B. Brimhall, and
     R. T. Jones. *N. Engl. J. Med. 280*:739-745 (1969).

154. E. Ahern, V. Ahern, T. Hilton, G. R. Serjeant, B. E. Serjeant,
     M. Seakins, A. Lang, A. Middleton, and H. Lehmann. *FEBS
     Letters 69*:99-102 (1976).

155. G. D. Efremov, T. H. J. Huisman, M. Stanulovic, M. Zurovec,
     H. Duma, J. B. Wilson, and V. Jeremic. *Scand. J. Haemat. 13*:
     48-60 (1974).

156. S. Shibata, T. Miyaji, I. Iuchi, Y. Ohba, and K. Yamamoto.
     J. Biochem. 63:193-198 (1968).

157. M. Aksoy, S. Erdem, G. D. Efremov, J. B. Wilson, T. H. J.
     Huisman, W. A. Schroeder, J. R. Shelton, J. B. Shelton,
     O. N. Ulitin, and A. Müftüoglu. *J. Clin. Invest. 51*:2380-
     2387 (1972).

158. R. Finney, R. Casey, H. Lehmann, and W. Walker. *FEBS Letters
     60*:435-438 (1975).

159. N. B. Dobbs, Jr., J. W. Simmons, J. B. Wilson, and T. H. J. Huisman. *Biochim. Biophys. Acta 117*:492-494 (1966).

160. F. Taketa, Y. P. Huang, J. A. Libnoch, and B. H. Dessel. *Biochim. Biophys. Acta 400*:348-353 (1975).

161. S. H. Boyer, S. Charache, V. F. Fairbanks, J. E. Maldonado, A. Noyes, and E. E. Gayle. *J. Clin. Invest. 51*:666-676 (1972).

162. S. J. Zak, G. R. Geller, W. Krivit, D. Tukey, B. Brimhall, R. T. Jones, H. F. Bunn, and M. McCormack. *Br. J. Haematol. 33*:101-104 (1976).

163. D. R. Miller, R. I. Weed, G. Stamatoyannopoulos, and A. Yoshida. *Blood 38*:715-729 (1971).

164. P. R. Pedersen, P. R. McCurdy, R. N. Wrightstone, J. B. Wilson, L. L. Smith, and T. H. J. Huisman. *Blood 42*:771-781 (1973).

165. D. R. Miller, R. I. Weed, G. Stamatoyannopoulos, and A. Yoshida. *Blood 38*:715-729 (1971).

166. H. Wajcman, J. Pagnier, D. Labie, and P. Boivin. *Nouv. Rev. Fr. Hémat. 11*:317-330 (1971).

167. R. T. Jones, E. E. Osgood, B. Brimhall, and R. D. Koler. *J. Clin. Invest. 46*:1840-1847 (1967).

168. M. Nagai, M. Nishibu, Y. Sugita, Y. Yoneyama, R. T. Jones, and S. Gordon. *J. Biol. Chem. 250*:3169-3173 (1975).

169. C. S. Reed, R. Hampson, S. Gordon, R. T. Jones, M. J. Novy, B. Brimhall, M. J. Edwards, and R. D. Koler. *Blood 31*:623-632 (1968).

170. D. J. Weatherall, J. B. Clegg, S. T. Callender, R. M. G. Wells, R. E. Gale, E. R. Huehns, M. F. Perutz, G. Viggiano, and C. Ho. *Br. J. Haematol. 35*:177-191 (1977).

171. M. J. Mant, M. L. Salkie, N. Cope, F. Appling, K. Bolch, M. Jayalakshmi, M. Gravely, J. B. Wilson, and T. H. J. Huisman. *Hemoglobin 1*:183-194 (1976).

172. R. T. Jones, B. Brimhall, and G. Gray. *Hemoglobin 1*:171-182 (1976).

173. J. Bonaventura and A. Riggs. *J. Biol. Chem. 243*:980-991 (1968).

174. R. L. Nagel, J. Lynfield, J. Johnson, L. Landau, R. M. Bookchin, and M. B. Harris. *N. Engl. J. Med. 295*:125-130 (1976).

175. G. D. Efremov, E. Stojmirovic, H. L. Lam, J. B. Wilson, and T. H. J. Huisman. *Hemoglobin 2*:75-77 (1978).

176. G. D. Efremov, T. H. J. Huisman, L. L. Smith, J. B. Wilson, J. L. Kitchens, R. N. Wrightstone, and H. R. Adams. *J. Biol. Chem. 244*:6105-6116 (1969).

177. T. Wilkinson, C. G. Chua, R. W. Carrell, H. Robin, T. Exner, K. M. Lee, and H. Kronenberg. *Biochim. Biophys. Acta 393*: 195-200 (1975).

178. E. Kohne, H. P. Kley, E. Kleihauer, H. Versmold, H. C. Benöhr, and G. Braunitzer. *FEBS Letters 64*:443-447 (1976).

179. T. Imamura, S. Fujita, Y. Ohta, M. Hanada, and T. Yanase. *J. Clin. Invest. 48*:2341-2348 (1969).

180. M. A. R. King, B. G. Wiltshire, H. Lehmann, and H. Morimoto. *Br. J. Haematol. 22*:125-134 (1972).

181. H. M. Ranney, A. S. Jacobs, and R. L. Nagel. *Nature 213*:876-878 (1967).

182. E. E. Di Iorio, K. H. Winterhalter, K. Wilson, A. Rosenmund, and H. R. Marti. *Blut 31*:61-68 (1975).

183. R. G. Schneider, J. B. Alperin, B. Brimhall, and R. T. Jones. *J. Lab. Clin. Med. 73*:616-622 (1969).

184. W. F. Moo-Penn, K. C. Bechtel, M. H. Johnson, D. L. Jue, B. L. Therrell, B. Y. Morrison, and R. M. Schmidt. *Biochim. Biophys. Acta 453*:472-477 (1976).

185. R. G. Schneider, N. L. Berkman, B. Brimhall, and R. T. Jones. *Biochim. Biophys. Acta 453*:478-483 (1976).

186. M. A. F. El-Hazmi and H. Lehmann. *Hemoglobin 1*:59-74 (1976-77).

187. P. F. Milner, C. Miller, R. Grey, M. Seakins, W. W. De Jong, and L. N. Went. *N. Engl. J. Med. 283*:1417-1425 (1970).

188. C. Altay, N. Altinoz, J. B. Wilson, K. C. Bolch, and T. H. J. Huisman. *Biochim. Biophys. Acta 434*:1-3 (1976).

189. G. Martinez, F. Lima, and B. Colombo. *Biochim. Biophys. Acta 491*:1-6 (1977).

190. R. Q. Blackwell, H.-J. Yang, and C.-C. Wang. *Biochim. Biophys. Acta 194*:1-5 (1969).

191. R. Q. Blackwell, P. R. McCurdy, C.-S. Liu, C.-L. Wang, and J. T.-H. Huang. *Vox Sang. 28*:50-56 (1975).

192. M. Hubbard, J. B. Wilson, R. N. Wrightstone, G. D. Efremov, and T. H. J. Huisman. *Acta Haemat. 54*:53-58 (1975).

193. B. Ringelhann, F. I. D. Konotey-Ahulu, N. C. Talapatra, F. K. Nkrumah, B. G. Wiltshire, and H. Lehmann. *Acta Haemat. 45*: 250-258 (1971).

194. W. Moo-Penn, D. Jue, B. George, N. Ramsey, and R. M. Schmidt. *Am. J. Clin. Path. 63*:87-90 (1975).

195. G. H. Bare, J. O. Alben, P. A. Bromberg, R. T. Jones, B. Brimhall, and F. Padilla. *J. Biol. Chem. 249*:773-779 (1974).

196. M. Jenson, F. A. Oski, D. G. Nathan, and H. F. Bunn. *J. Clin. Invest.* *55*:469-477 (1975).

197. S. J. Zak, B. Brimhall, R. T. Jones, and M. E. Kaplan. *Blood* *44*:543-549 (1974).

198. H. B. Kleckner, J. B. Wilson, J. G. Lindeman, P. D. Stevens, G. Niazi, E. Hunter, C. J. Chen, and T. H. J. Huisman. *Biochim. Biophys. Acta* *400*:343-347 (1975).

199. J. W. Adamson, J. T. Parer, and G. Stamatoyannopoulos. *J. Clin. Invest.* *48*:1376-1386 (1969).

200. J. W. Adamson, A. Hayashi, G. Stamatoyannopoulos, and W. F. Burger. *J. Clin. Invest.* *51*:2883-2888 (1972).

201. H. F. Bunn, T. B. Bradley, W. E. Davis, J. W. Drysdale, J. F. Burke, W. S. Beck, and M. B. Laver. *J. Clin. Invest.* *51*:2299-2309 (1972).

202. H. Wajcman, J. V. Kilmartin, A. Najman, and D. Labie. *Biochim. Biophys. Acta* *400*:354-364 (1975).

203. F. Vella, S. C. Wong, J. B. Wilson, and T. H. J. Huisman. *Can. J. Biochem.* *50*:841-843 (1972).

204. H. M. Ranney, A. S. Jacobs, B. Ramot, and T. B. Bradley, Jr. *J. Clin. Invest.* *48*:2057-2062 (1969).

205. W. W. De Jong and L. N. Went. *Human Heredity* *24*:32-39 (1974).

206. F. W. Boerma, J. Nijboer, F. Vella, S. C. Wong, and T. H. J. Huisman. *Clin. Chim. Acta* *55*:49-55 (1974).

207. R. T. Jones, B. Brimhall, and T. H. J. Huisman. *J. Biol. Chem.* *242*:5141-5145 (1967).

208. R. F. Rieder, J. B. Clegg, H. J. Weiss, N. P. Christy, and R. Rabinowitz. *Biochim. Biophys. Acta* *439*:501-504 (1976).

209. R. S. Sharma, D. L. Harding, S. C. Wong, J. B. Wilson, M. E. Gravely, and T. H. J. Huisman. *Biochim. Biophys. Acta* *359*: 233-235 (1974).

210. R. Alberti, L. Tentori, M. Marinucci, and V. Borghesi. *Hemoglobin* *2*:171-174 (1978).

211. L. E. Lie-Injo, W. Pribadi, F. W. Boerma, G. D. Efremov, J. B. Wilson, C. A. Reynolds, and T. H. J. Huisman. *Biochim. Biophys. Acta* *229*:335-342 (1971).

212. R. S. Sharma, L. Williams, J. B. Wilson, and T. H. J. Huisman. *Biochim. Biophys. Acta* *393*:379-382 (1975).

213. W. W. De Jong and L. F. Bernini. *Nature* *219*:1360-1362 (1968).

214. T. H. J. Huisman, C. A. Reynolds, A. M. Dozy, and J. B. Wilson. *Biochim. Biophys. Acta* *175*:223-225 (1969).

215. L. E. Lie-Injo, H. Kamuzora, and H. Lehmann. *J. Med. Genet.* *11*:25-30 (1974).

216. E. J. Ahern, B. G. Wiltshire, and H. Lehmann. *Biochim. Biophys. Acta 271*:61-64 (1972).

217. R. G. Schneider and R. T. Jones. *Science 148*:240-242 (1975).

218. R. W. Carrell, M. C. Owen, R. Anderson, and E. Berry. *Biochim. Biophys. Acta 365*:323-327 (1974).

219. S. O. Brennan, M. B. Smith, and R. W. Carrell. *Biochim. Biophys. Acta 490*:452-455 (1977).

220. L. E. Lie-Injo, B. G. Wiltshire, and H. Lehmann. *Biochim. Biophys. Acta 322*:224-230 (1973).

221. E. J. Ahern, R. T. Jones, B. Brimhall, and R. H. Gray. *Br. J. Haematol. 18*:369-375 (1970).

222. E. Ahern, W. Holder, V. Ahern, G. R. Serjeant, B. E. Serjeant, M. Forbes, B. Brimhall, and R. T. Jones. *Biochim. Biophys. Acta 393*:188-194 (1975).

223. M. N. Cauchi, J. B. Clegg, and D. J. Weatherall. *Nature 223*: 311-313 (1969).

224. G. Altay, F. Garver, W. H. Bannister, J. L. Grech, A. Felice, and T. H. J. Huisman. *Biochem. Genet. 15*:915-923 (1977).

225. E. J. Ahern, V. Ahern, B. G. Wiltshire, and H. Lehmann. *Biochim. Biophys. Acta 303*:242-245 (1973).

226. B. Brimhall, T. S. Vedvick, R. T. Jones, E. Ahern, E. Palomino, and V. Ahern. *Br. J. Haematol. 27*:313-318 (1974).

227. L. E. Lie-Injo, C. G. Lopez, and M. Lopez. *Acta Haemat. 46*: 106-120 (1971).

228. J. B. Clegg, D. J. Weatherall, and P. F. Milner. *Nature 234*: 337-340 (1971).

229. J. B. Clegg, D. J. Weatherall, I. Contopolou-Griva, K. Caroutsos, P. Poungouras, and H. Tsevrenis. *Nature 251*:245-247 (1974).

230. W. W. De Jong, P. M. Khan, and L. F. Bernini. *Am. J. Hum. Genet. 27*:81-90 (1975).

231. M. Seid-Akhavan, W. P. Winter, R. K. Abramson, and D. L. Rucknagel. *Proc. Natl. Acad. Sci. USA 73*:882-886 (1976).

232. H. F. Bunn, G. J. Schmidt, D. N. Haney, and R. G. Dluhy. *Proc. Natl. Acad. Sci. USA 72*:3609-3613 (1975).

233. T. H. J. Huisman, J. B. Wilson, M. Gravely, and M. Hubbard. *Proc. Natl. Acad. Sci. USA 71*:3270-3273 (1974).

234. E. J. Ahern, V. N. Ahern, G. H. Aarons, R. T. Jones, and B. Brimhall. *Blood 40*:246-256 (1972).

235. G. D. Efremov, R. Rudivic, G. A. Niazi, E. Hunter, Jr., T. H. J. Huisman, and W. A. Schroeder. *Scand. J. Haemat. 16*:81-89 (1976).

236. J. Barnabas and C. J. Muller. *Nature 194*:931-932 (1962).

237. W. A. Schroeder, T. H. J. Huisman, C. Hyman, J. R. Shelton, and G. Apell. *Biochem. Genet. 10*:135-147 (1973).

238. W. F. Moo-Penn, K. C. Bechtel, and B. L. Therrell, Jr. *Hemoglobin 2*:65-69 (1978).

239. G. R. Honig, M. Shamsuddin, R. G. Mason, and L. N. Vida. *Proc. Natl. Acad. Sci. USA 75*:1475-1479 (1978).

240. T. H. J. Huisman, R. N. Wrightstone, J. B. Wilson, W. A. Schroeder, and A. G. Kendall. *Arch. Biochem. Biophys. 153*: 850-853 (1972).

241. A. G. Kendall, P. J. Ojwang, W. A. Schroeder, and T. H. J. Huisman. *Am. J. Hum. Genet. 25*:548-563 (1973).

242. R. F. Rieder and G. W. James, III. *J. Clin. Invest. 54*:948-956 (1974).

243. M. Cohen-Solal, Y. Blouquit, M. C. Garel, J. Thillet, L. Gaillard, R. Creyssel, A. Gibaud, and J. Rosa. *Biochim. Biophys. Acta 351*:306-316 (1974).

244. R. T. Jones, B. Brimhall, T. H. J. Huisman, E. Kleihauer, and K. Betke. *Science 154*:1024-1027 (1966).

245. H. Wajcman, D. Labie, and G. Schapira. *Biochim. Biophys. Acta 295*:495-504 (1973).

246. R. F. Rieder. *J. Clin. Invest. 50*:388-400 (1971).

247. J. Murari, L. L. Smith, J. B. Wilson, R. G. Schneider, and T. H. J. Huisman. *Hemoglobin 1*:267-282 (1977).

248. C. L. Lutcher, J. B. Wilson, M. E. Gravely, P. D. Stevens, C. J. Chen, J. G. Lindeman, S. C. Wong, A. Miller, M. Gottlieb, and T. H. J. Huisman. *Blood 47*:99-112 (1976).

249. W. Moo-Penn, K. Bechtel, D. Jue, M. S. Chan, G. Hopkins, N. J. Schneider, J. Wright, and R. M. Schmidt. *Blood 46*: 363-367 (1975).

250. R. M. Winslow, M. L. Swenberg, E. Gross, P. A. Chervenick, R. R. Buchman, and W. F. Anderson. *J. Clin. Invest. 57*:772-781 (1976).